# FUSION ENERGY

# FUSION ENERGY

*Introduction and Magnetic Confinement*

**SUBHAAN AKHTAR**

Fusion Energy
Subhaan Akhtar

Published by White Falcon Publishing
Chandigarh, India


First Edition, 2023

Cover/Interior designed by Subhaan Akhtar



ISBN - 978-81-19510-28-3

*"I would like nuclear fusion to become a practical power source. It would provide an inexhaustible supply of energy, without pollution or global warming."*

**STEPHEN HAWKING**

## Table of Contents

# CH 1: The Equivalence of Mass and Energy

“E=mc2. Energy equals mass times the speed of light squared”

- Albert Einstein

To understand the relationship between mass and energy, it is necessary to look at two principles that have held a high place in pre-relativity physics, mainly the conservation of energy and the conservation of mass. To illustrate the former, consider a pendulum whose mass swings between points A and B. The mass m is higher by the amount h than it is at c, the lowest point of the path the mass travels (see fig. 1).

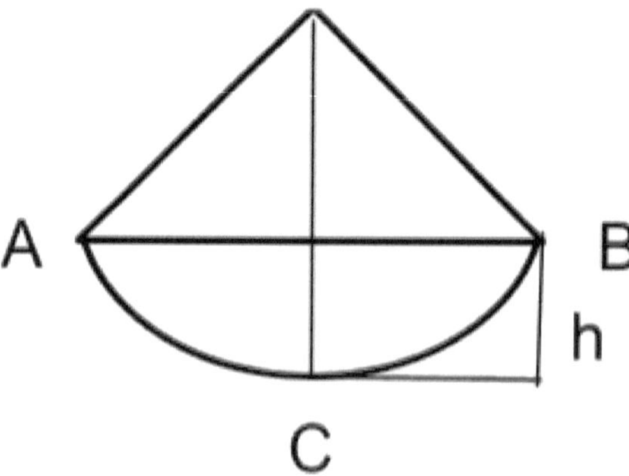


Figure 1: Diagram of a pendulum

But, at C, the lifting height has completely disappeared, and instead, the mass has velocity v. This relationship can be expressed as

$$mgh = (m/2)\, v^2$$

where g is representing gravitational acceleration, showing how velocity can be entirely converted into lifting height, and vice versa. What's fascinating about this relationship is that it doesn't change when the form of the path through which the pendulum is travelling changes, or when the length of the pendulum changes. No. Rather, something remains constant during this process, and that thing is energy. At A and B, there is potential energy- energy the mass has by virtue of its position. Similarly, at C, there is kinetic energy- the energy of motion. Hence, if v is understood to represent velocity at that point in the pendulum's path, and h the height above C, the sum

$$mgh + m (v^2/2)$$

must have the same value for any position on the pendulum. And since this is found to be the case, this principle can be generalised to get the law of the conservation of mechanical energy. The answer to the effects of friction can be found in the study of heat phenomena. When it was discovered that for any amount of heat produced by friction, an exactly proportional amount of energy has to be expended, the principle of the equivalence of heat and work was discovered. This is seen in the pendulum, as friction gradually converts the mechanical energy into heat. Thus, the principles of mechanical and thermal energy are merged into one. This then begs the question: Can the principle of conservation be extended to electromagnetic and chemical processes, and applied to all fields? Overall, it appeared as though the physical system comprised a sum total of energies that remained constant in spite of any changes that might occur. It only makes sense then, that this principle would apply to mass. Mass, both inert and heavy, as different as these two definitions are, appeared to be an unvarying and essential quality of matter. Combining into chemical compounds, or heating, melting, and vaporisation would not change the total mass, and hence, the principle of the conservation of mass was arrived at, namely, that masses remain unchanged under any physical or chemical changes.

However, this principle was abandoned a century ago as it proved inadequate in the face of the special theory of relativity. Rather, it was merged with the principle of the conservation of energy, in such a fashion as the principle had been extended to other processes and qualities of matter. The equivalence of mass end energy is customarily (though somewhat inexactly) given by the formula

$$E=mc^2$$

where E is the energy contained in a stationary body and m its mass, and c, the velocity of light in a vacuum. From this equation, it is made evident that since mass is multiplied by the enormous square of the speed of light, a small amount of mass can be converted into a large amount of energy. It is this principle that makes nuclear fission a viable source of energy, and fusion even more so. It should be noted that reversing the relation would result in an increase in mass too small, as is evident when the equation is rearranged as

$$m=E/c^2$$

where the occurrence of the enormous factor $c^2$ in the denominator necessitates the energy to be a large amount for even a small increase in mass- hence heating a mass, say, 10 degrees, won't result in an increase that can be measured even with the most sensitive balance.

Considering this important relation at the heart of nuclear reactions, both fission and fusion and particularly in light of the latter, it would be useful to represent it in a manner that delineates the steps and conditions for fusion, eventually allowing for a determination of the factors which allow for confinement and ignition. Since it is the case that the strong force- the force that holds nuclei together in spite of the proximity of protons to each other- has a very short range ($10^{-15}$ metres), in most instances of collision the particles repel each other. This is due to the fact that the strong force involves the exchange of particles. Specifically the exchange of gluons between quarks to form protons and neutrons, and the exchange of mesons between protons and neutrons to form atomic nuclei. Keep in mind that this force- called the Coulomb force, increases inversely to the square of the distance, and hence, nearing the range of the strong force is very difficult to overcome. This is a part of the reason why fusion, if it is to generate energy, necessitates extremely high temperatures and pressures, along with long confinement times (the interaction between these properties of any fusion plasma is the subject of the Lawson Criterion). In light of this, it would be useful to add an expression for the energy $E_{in}$ required to facilitate a reaction. Doing so would provide us, with mass M and energy E the following:

$$\mathbf{Ein + Min \rightarrow Eout + Mout}$$

Evidently, the occurrence of a matter-energy transformation (see fig. 2), which can constitute a wide array of reaction and ignition methods will be shown, requires the following conditions to be satisfied:

**Mout < Min & Ein > Ein**

Figure 2: Matter-Energy Transformation

Later the interaction between two identical as well as different types of particles will be considered, but for now, it would be useful to modify the relations mentioned above so as to arrive at those which shed some light on the preconditions for fusion. Starting with energy: Since any method of bringing two particles close together would involve accelerating them, and hence the energy involved is kinetic, the energy added can be denoted by $E_k$. At the same time, the mass-energy can be denoted by $E_r$. Evidently, such distinctions are useful when mass and energy are interchangeable. Moreover, considering the fact that total energy E* of an ensemble of particles remains constant, ie. $E^*_{Before} = E^*_{After}$ (obeying the first law of thermodynamics), to account for the added kinetic energy the original equation of mass and energy can be written as

$$E^* = \Sigma (E_{kj} + E_{rj}) = \Sigma (E_{k,j} + m_j c^2)$$

## Fusion Fuel Interaction

In its current state this equation provides no insight into how fusion fuels will interact. Hence, it will be necessary to include particles a, b, d, and e, each of which is interacting with the others like so:

$$a + b \rightarrow d + e$$

Making the value of E*, with both kinetic and mass energy considered for all 4 particles, now given by:

$$(E_{ka}+m_a c^2) + (E_{kb} + m_b c^2) = (E_{kd}+m_d c^2) + (E_{ke} + m_e c^2)$$

For proof of conservation of energy, look no further than the rearrangement of these terms, which shows that the value of the initial energy subtracted from the final energy is the same as that of the final mass subtracted from the initial mass (this is because of the fact that the reaction resulted in less mass and more energy than before):

$$(E_{kd} + E_{ke}) - (E_{ka} + E_{kb}) = [(m_a + m_b) - (m_d + m_e)]c^2$$

It is possible to test the above statement taking initial values of E*=50 and M=50, with final values of E*=80 and M=20. Subtracting the initial mass from the final, as well as subtracting the final energy from the initial, yields the same value of 30 (in terms of mass and energy respectively), thus, proving that energy is conserved. Finally, using the symbols a, b, d, and e which represent different types of particles, it is possible to represent the fact that, just as the number of atoms is conserved in a chemical reaction, the number of nucleons is as well as such:

$$A_a + A_b = A_d + A_e$$

$$Z_a + Z_b = Z_d + Z_e$$

Here A is the number of nucleons and Z the number of which are protons such that with a nuclear species named X they are given as follows:

$$^{Aa}{}_{Za}X_a + {}^{Ab}{}_{Zb}X_b \rightarrow {}^{Ad}{}_{Zd}X_d + {}^{Ae}{}_{Ze}X_e$$

This can also be characterised kinematically, obeying the law of the conservation of momentum through the fact that the mass multiplied by the velocity of end product d is equal to that of end product e ($m_dv_d = m_ev_e$) (see fig. 3).

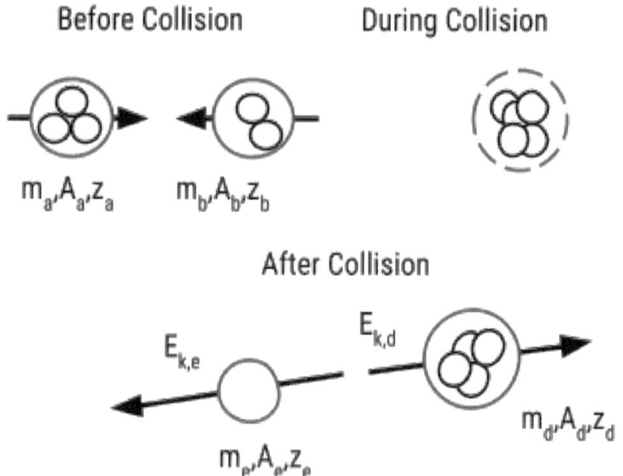


Figure 3: Kinematic characterisation of fuels a, b, d, & e

Replacing a, b, d, and e with the reactants of a fusion reaction. Since two isotopes of hydrogen: deuterium and tritium (the former with one additional neutron and the latter with two) comprise the most popular selection, they will be considered so that that a=d, b=t, d= a (helium nucleus, called an alpha particle), and e= n (neutron) as such:

$$\mathbf{d + t \rightarrow a + n + 17.6\ MeV}$$

In this case 3.5 MeV of the energy released is in the alpha particle while 14.1 MeV is in the neutron. This becomes relevant for achieving self-sustaining conditions in a reactor, as the alpha particle wouldn't be affected by the magnetic field, and hence contribute its energy towards heating the plasma. Moreover, that the neutron can escape the field means that its kinetic energy can be used for the production of electricity.

Such a process involves capturing neutrons in a material called a "blanket", thus, allowing it to heat up. This heat is then transferred to a coolant such as water, helium, or liquid salts, and is then converted into electricity using traditional methods- producing steam in a heat exchanger which drives a turbine connected to an electrical generator. Another common reaction is that between two deuterons, the so-called d-d reaction. Such a reaction is possible under 2 distinct channels. Both of these channels occur with almost equal probabilities under identical reaction conditions:

$$\mathbf{d + d \rightarrow p + t + 4.1\ MeV}$$
$$\mathbf{d + d \rightarrow n + h + 3.2\ MeV}$$

Evidently more reactions are possible under these conditions due to the fact that the tritium product can fuse with deuterium. Moreover, d-p reactions are also possible, their combination forming a helium-3 nucleus and the release of a high-energy gamma photon (γ). Another possible reaction is one that results in the formation of helium-3 and a high-energy neutron. Both of these reactions can be expressed as follows:

$$d + p \rightarrow {}^{3}He + \gamma$$
$$d + p \rightarrow {}^{3}He + n$$

Keep in mind the fact that understanding how these reactions would occur under reactor conditions necessitates an understanding of the probability of each reaction. Such probabilities are determined by which nuclei interact. For instance, any reaction with a hydrogen nucleus (a proton), is highly unlikely due to the small reaction cross-section and high electrostatic repulsion that can more easily accelerate a light particle such as a proton.

In addition, such analysis yields another important practical consideration- that d-d reactions aren't aneutronic. That is, they face the same limitations as d-t fusion when it comes to the need to ensure the materials of the wall don't mix with the plasma, thus cooling it. Moreover, it means that the reactor walls would require to be replaced frequently due to the buildup of holes that make them unsuitable for maintaining a vacuum. It is for these reasons that the concept only remains as a means to conduct experimental fusion research in a manner that is inexpensive.

## Basic Parameters- Temperature, Pressure, Time

Even with as brief of an analysis of the implications of the relationship between matter and energy for fusion reactions as provided, the reader notices certain conditions necessary for controlled, sustained, fusion. The role of temperature is what first comes to mind. It is necessary that particles, regardless of the means by which they are accelerated, are able to do so at a velocity that allows them to overcome the Coulomb barrier. Hence, most fusion reactor concepts first aim to create the high-temperature conditions that are necessary for nuclei to fuse- in the process creating the state of matter in which electrons are stripped from their nucleus called plasma.

However, at scales such as those considered for fusion plasmas (as well as for others, although not to have any great effect on results), it is important to note that temperature refers to the average particle kinetic energy of the particles. Consequently, the notion of temperature will be abandoned in favour of electron volts (eV), a measure of the amount of kinetic energy gained by a single electron accelerating from rest through an electrical potential difference of a single volt in a vacuum. The conversion factor of the two (with temperature measured in Kelvin) is 11606, in the sense that 1 eV corresponds to 11606 K and 1 keV corresponds to 11.6 million degrees Kelvin.

Nevertheless, it would appear rather ostensibly that higher temperatures would to some degree be counterproductive to the effort of sustaining controlled fusion in a manner that is net positive in light of the other important factor that would seem to determine the effectiveness of plasma confinement- pressure. This is due to the fact that there is no way to confine a hot plasma using materials, as they would melt and cool down the plasma, thus necessitating the use of electrostatic fields, lasers, liquid metals, or magnetic fields (among other methods). However, the effectiveness of such methods decreases as individual particles possess the kinetic energy to escape confinement, thus, reducing pressure. This is, however, a rather incomplete analysis, as it fails to account for the fact that each is responsible for giving rise to conditions in the plasma which disrupt the other as well as confinement (known as instabilities). Later chapters when accounting for the magnetohydrodynamic properties of confined plasmas will be able to provide a more coherent argument- although arriving at the same conclusion.

## Better than the Sun

While a fusion reactor is often referred to as a 'sun on earth', and the use of one is considered 'bottling a star', it should be noted that what is needed to sustain reactions that can efficiently generate energy are conditions different from those of the sun. Considering that hydrogen atoms in the sun's interior- where the temperature is about 1keV- live for a million years before they fuse into helium and release energy, a working reactor- and here working refers to one that not only sustains reactions but does so only using so much power that the reactor has an efficiency of 10% or above (the minimum efficiency for the construction and operation of reactors to be viable)- needs to increase the temperature to speed up the process, while at the same time increase the inertia in head-on collisions with heavier particles: the sun, by virtue of its gravitational pull, has circumvented to some pes such as deuterium and tritium to overcome the electrostatic repulsion of ions with like charges, a problem, the sun, by virtue of the enormous magnitude extent (nevertheless remaining inefficient). Hence, the challenge to create a viable fusion reactor is justified, and seemingly so is the cynicism of the critics of fusion.

## A Note to the Reader

Finally, before proceeding into the various aspects of fusion reactors, the reader should have come to the realisation that the challenge of fusion is one that lies in the complexity associated with the interaction of multiple factors simultaneously. And while this has always made the methods of scientific investigation struggle against the behaviour of fluids, it is now more pronounced in fusion reactors as a result of the fact that plasmas conduct and create electromagnetic fields. This book, if anything, aims to extract from the wealth of fusion literature that which would allow for the understanding of such interactions, mainly a set of conclusions which reflect some fundamental truth of their nature akin to the significance of the relation of mass and energy. If any reader is disillusioned by this fact, then they should principally consider the fact that currently, in spite of the prevalence of numerous renewable sources of energy such as wind, solar, and hydropower, as well as others such as nuclear fission, most of the energy humans consume comes from the burning of fossil fuels such as coal and natural gas. And while the estimates of how long before the climate crisis reaches a point at which there is no opportunity for recompense vary greatly- ranging from 12 to over 50 years), there is no doubt that a change in the energy landscape is necessary for humanity to meet the needs of its future generations in a manner that doesn't lead to ecological catastrophe. The potential fusion energy posses, in both the short and long term, is enough to justify the attempt to develop it. Finally, the book will often include equations with asterisks to denote the fact that the magnitudes to which they are attached are referring to the entire reaction volume as opposed to the respective density expressions.

# CH 2: Controlled Fusion: The Future of Energy

"In my mind, (fusion) ranks with the original gift of fire, back in the mists of prehistory."

- Ben Bova

Before any attempt can be made to arrive at a set of fundamental principles both across all reactor concepts and pertaining to specific ones by virtue of revealing some underlying scientific fact, it is first necessary to summarize the current understanding of the conditions essential for harnessing fusion reactions of the sort described in the previous chapter for the purpose of generating energy. Such an attempt is here made, and it comprises both a chronological description of the advancements made in the development, construction, and operation of controlled fusion reactions, as well as the consequent advancements to scientific understanding. The latter in this case is explored in terms of both the energy transformations involved in the formation of heavy nuclei (nuclear energetics), as well as the preconditions for achieving a state at which point energy release from fusion reactions can become a self-sustaining process (ignition). Note that these considerations don't pertain to any given reactor concept, but rather establish the principles upon which those concepts find their scientific underpinnings. Thus, this section is in some semblance the starting point for the desired outcome of this book, and hence, is here to provide the reader with the context in which to understand any conclusions arrived at later.

# History of Fusion Reactors

## Accelerator-Based Experimentation

While the promise of fusion is obvious today, during the early twentieth century, it was only beginning to be explored. It wasn't until fusion was first achieved on earth by Ernest Rutherford, Mark Oliphant, and Paul Harteck in 1934 that the premise of nuclear reactions as a means of generating energy was, and could, be taken seriously. The challenges intrinsic to fusion reactors in regards to producing more energy than is consumed were made evident by an experiment in 1934, in which a particle accelerator was used to shoot deuterium nuclei into metal foil containing a variety of nuclei. This experiment, which enabled Rutherford, Oliphant, and Harteck to measure the nuclear cross-sections of various reactions, yielded two primary findings: first, the reaction that occurred with the lowest energy as compared to those tested was the deuterium-deuterium one- peaking at around 100 keV (100,000 electron volts), and second, that most accelerated particles when colliding with the fuel will not fuse with it, but rather scatter off each other, and thus, lower the reaction rate and lead to energy losses via the kinetic energy transferred to the fuels. From these, the corollary that accelerator-based fusion is not practical due to the small size of the reactor cross-section- the power input even with deuterium-deuterium reactions not being efficient enough to sustain energy losses of such a magnitude- can be deduced. The problem of producing a net-positive energy output from fusion can be solved by raising the atoms are continually colliding at high speed such that the rate that energy is released by the reaction is high enough to heat the surrounding fuel to the degree that temperate can be maintained against losses to the environment, thus, enabling the reactions to continue. It was the great

Enrico Fermi who calculated the temperature at which point the rate energy is lost to the environment is high enough that the environment itself can heat the fuel rapidly enough to maintain its temperature against the rate that it gives off energy to its surroundings, and so reactions can occur in a continuous manner, to be around 50,000,000K.

## Proyecto Huemul

While the United States was still a year away from creating the H-bomb in a race against the Soviet Union (who didn't achieve this feat until 1953), and both countries or in fact, the rest of the world, hadn't considered the possibility of using light atoms for powering the world's growing energy demands, Argentina, a then largely immigrant rural nation of barely 16 million inhabitants, was able to create a reactor, that, according to Argentinian president Juan Perón, produced "the controlled liberation of atomic energy," through hydrogen, rather than uranium. That the discovery would bring "a greatness which today we cannot imagine", and was "transcendental for the future life" according to Perón only fuelled the growing interest in the reactor and the Austrian-born scientist and recent immigrant to Argentina named Ronald Richter, who claimed to have achieved a "net positive result' with hydrogen on 16'th February 1951. This sensational news was obviously false, and it didn't take the scientific community long to dismiss the Perón-Richter claim on the basis of a simple calculation that showed that the experimental setup- involving hydrogen fed into an electric arc- was incapable of producing enough energy to heat the fusion fuel to the necessary temperatures for fusion to occur. For having misled Perón, resulting in the president being embarrassed on an international scene, Richter was jailed. For any readers interested, an article about scientific research in the 1950s (Jan 2003, Vol. 56, N°1), by physicist Juan Roederer provides an interesting viewpoint into Richter's work- especially in the conversation it triggered between the author and physicist and fusion activist Friedwardt Winterberg regarding Richter' novel method of ion-acoustic plasma heating that will be considered in chapter 2. The reason this incident is relevant to the history of fusion is that the widespread news coverage meant politicians were made aware of fusion. This was a pivotal point for research, so much so that according to the International Thermonuclear Experimental Reactor (ITER) Newsline, "the unveiling of Proyecto Huemul triggered what is now recognised as the first decisive step into serious research in controlled fusion." The histories of all other reactors stem from this point, and hence, the implications of this event will be explored in the subsequent chapters concerning individual reactor concepts.

## Conditions for Viability

Subsequent sections will consider what decades of fusion research have found to be the conditions which any fusion reactor must satisfy in order to produce net-positive energy. Discussions of burn processes and reactor energetics are relegated to a later part of the chapter, however, and for the time being the considerations don't address fundamental aspects of plasma behaviour, acting more as indicators of whether any mode of operation is unviable on the basis of the simplest of facts which can be gathered about it. Thus, the following points are almost universally applicable, and any discussion of viability must principally be addressed in terms of them before a concept can be given any more attention than is due.

### Beta Parameter

When it comes to finding conditions that a fusion reactor must specify, there are some which pertain specifically to certain reactor types. Such a parameter for inertial confinement reactors will be considered in its respective chapter, though since various reactor concepts are predicated on the same principles of magnetic confinement, one that pertains to it is presented here.

The Beta Parameter refers to the ratio of kinetic particle pressure ($P_{tin}$) to magnetic pressure ($P_{mag}$). Both of these values have been discussed above, and they are respectively by

$$P_{tin} = N_i kT_i + N_e kT_e$$

and

$$P_{mag} = B^2/2\mu$$

The parameter offers a way in which to compare states of confinement based on how effectively the magnetic field constrains the thermal motion of the plasma particles. The goal of a system would be to have a low beta, as this would imply that it is able to confine a high-temperature plasma using weaker magnetic fields. And while meeting such a condition is useless if pressure and confinement time requirements or any other factor is not met as well, it acts as a useful metric for determining both the efficiency as well as the improvements that need to be made to a reactor. For the purposes of this book, the use of such a parameter will reveal flaws intrinsic to each form of magnetic confinement which restrict the ratio of kinetic particle pressure to magnetic pressure to certain limits.

## Fusion Energy Production Possible

Another important consideration is how much a reactor can theoretically produce given the type of fuel, its quantity, and the conditions to which it is exposed. More specifically, the fusion energy production possible in a deuterium-tritium plasma can be assessed with

$$P_{fu}=N_d N_t < \sigma v_{>dt} Q_{dt}$$

Note that this assumes the following:

$$N_d = N_t = N_i/z$$

$$N_i = N_e$$

$$T_i = T_e$$

Thus another equation can be framed that relates fusion power density and magnetic pressure, giving the magnetic pressure limited fusion power density as follows:

$$P_{fu,\ mag} = [(\beta^2_{max} B^4)/\ \ ] \times [(\sigma v_{>dt})/(T^2 i)]\ Q_{d,t}$$

This implies that there is a unique correlation between fusion power density (W/m$^3$) and temperature. For different values of ion temperature (keV), there would evidently be values of $P_{fu,\ mag}$. With all the assumptions made above, the relationship between the two can be plotted on a graph as shown in fig. 4) The fact that the relationship between values is not linear, but rather, changes means that certain temperatures are more favourable than others. For instance, a decline in fusion power density is observed for a field strength of 5 and 10 Tesla beyond 40-50 MeV, and hence, it wouldn't be favourable to heat the plasma further. Such an outcome is visible across many parameters, indicating that a fundamental aspect of plasmas is their propensity to change their behaviour. And while it is possible to predict this behaviour, under the extreme conditions of a fusion reactor, the method by which to do so becomes unclear.

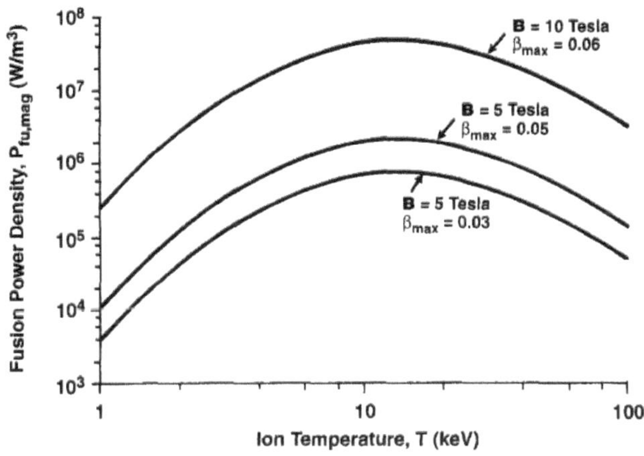


Figure 4: Pressure-limited fusion Power Density in a d-t fusion plasma

Such unique interaction can be seen principally in terms of individual trajectories- the motion of single particles with a given energy under the force of a magnetic field. Put many of these particles together, however, and the resulting behaviour is unlike that previously. For instance, a particle will move between field lines when near others, and its tendency to do so will change with the kinetic energy of itself along the field line as well as that of the particle that is responsible for the change. Evidently, they would both experience an increase in kinetic energy together, so for each increment, their interaction is altered. Now consider the case of millions of particles, all interacting with each other through their charge and kinetic energy. Thus, understanding their behaviour collectively would be necessary.

However, the fact that these two views characterise plasmas in such opposing ways becomes apparent when dealing with fusion plasmas. To illustrate, take for example the case of a magnetic field that is not homogenous. The trajectory of individual particles will evidently be altered, though the occurrence of this between the areas of varied field strength will result in the disruption being compounded. Such is only one example of the properties of plasma that make it prone to variations as seen in the graph, called an instability. Developing viable fusion reactors requires these to not only be accounted for, but exploited.

## Commercial Viability

Finally there is the practical consideration of whether a reactor can be utilised to generate electricity even if it produces more energy than it consumes. Reactors consist of large, heavy pieces, and are difficult to build. Moreover, most require some form of maintenance after being used. This includes replacing everything from small pellets to entire radioactive walls. However, so long as the net energy output is sufficient to justify the expenses associated with their construction and operation, fusion reactors have the potential to be as commercially viable as any other source of energy. Thus, the commercial viability will be considered in terms of the conditions necessary within the confined plasma for breakeven and ignition and will be the final condition considered for now.

Evidently, a fusion must have a net positive energy output ($E_{fu} >> E_{supply}$). And, since most magnetic confinement devices (both current and those planned for the future) operate in a pulsed mode- that is in intervals characterised by a burn time $\tau_b$, for the plasma containing a 50-50 split of deuterium and tritium, the fusion energy generated during said interval or time can be given by

$$E_{supply} = P_{fu} \times \tau_b = <\sigma v_{>dt} N_a N_t Q_{dt} \tau_b$$

In addition, considering a state in which the energy is perfectly coupled to an ensemble of deuterons, tritons, and electrons to which it is supplied, $E_{supply}$ can be given as

$$E_{supply} = 3/2\ N_d\ kT_d + 3/2N_t\ kT_e + 3/2N_t\ kT_e$$

Note that in both equations fusion power is assumed to be either constant throughout $\tau_b$ or to represent the average power over that interval. Neither would accurately depict the condition of the plasma, as changes in fusion power would be inevitable with shorter burn times, while the plasma would respond to its actual values, and not behave in accordance with the average value. Nevertheless, equating the ride sides of both the equations provides a useful definition of an ideal state of energy breakeven ($E_{fu} = E_{supply}$) as given by

$$<\sigma v_{>d\ t} N_d\ N_t\ Q_{dt}\ \tau_b = 3/2\ N_d\ kT_d + 3/2N_t\ kT_e + 3/2N_t\ kT_e$$

By imposing the conditions applied previously in the chapter (Nd = Nt = Ni/z, Ni = Ne) along with $T_d = T_t = T_e = T_i$, the expression can be simplified to now become

$$N_i\ \tau_b\ = 12kT\ /\ <\ \sigma v_{>dt} Q_{dt}$$

The product hare can be computed when kT= 12 keV because $Q_{dt}$ has a fixed value of 17.6 MeV while is a function of kT. Notwithstanding, this doesn't account for the inefficiencies associated with the process of energy conversion, nor does it take into consideration the energy losses that result from the electromagnetic radiation emissions as a result of the accelerated motion of charged particles. On the whole, such an ideal breakeven is an absolute minimum, and power losses of the sort mentioned above necessitate a higher calculation of the breakeven point, as well as a better means of analysis that is able to take into account the factors which ultimately affect energy efficiency.

Such methods of analysis are collectively known as the Lawson Criterion in recognition of the British engineer and physicist John D. Lawson, and they approach the problem by focussing on the fundamental factors that affect confinement: temperature, pressure, and confinement time. Due to the great value this evidently possesses in context to the aims of this book, the Lawson criterion will receive extensive treatment in the discussion of fusion reactor energetics later in the chapter.

## Fusion Burn

A burning plasma refers to one in which the energy released is sufficiently able to sustain itself, thus, heating the plasma without the need for external methods which contributes to making the process require more energy than is generated by the process. Here evidently reaction rates, as well as the time dependence of burn processes will determine how to create and sustain such operational conditions.

Considering the fact that deuterium-tritium fuel has received the most attention in this regard, it will first be extensively considered. However, subsequent sections will touch on identical fuel compositions such as deuterium-deuterium, as well as identify other factors which burn processes.

### Elementary and Comprehensive D-T Burn

As mentioned before, within a stable unit volume of interest, a d-t ion population will yield

$$d + t \rightarrow n + a$$

However, this doesn't describe the likelihood of such a reaction occurring under a given set of conditions. This so-called occurrence rate can be given by:

$$R_{dt}(r,t) = N_d(r,t)\ N_t\ (r,t) < \sigma v_{>dt}\ (r,t)$$

Based on this, it is now possible to determine, in the same way as was done earlier, the rate of energy-density release or fusion power density by including $Q_{dt}$=17.6 MeV and finding the value of $< \sigma v_{>dt}$ as a function of ion temperature with the following:

$$P_{dt}(r,t) = R_{dt}(r,t)\ Q_{dt} = N_d(r,t)\ N_t(r,t) < \sigma v_{>dt}\ (r,t)\ Q_{dt}$$

The fusion power in a unit volume, then, can be changed by varying both the ion temperature, as well as the fusile ion densities with time. Clearly, then, the previous assumptions of mean ion temperature or its remaining constant wouldn't accurately illustrate the burn processes in a fusion reactor. For instance, if the average particle kinetic energy of a group of particles for a certain part of the burn time is such that ignition conditions are achieved, then a decline later would not simply be compensating for that, as taking an average would seem to suggest. Moreover, as was shown above, increments in temperature correspond to fusion power density values in a manner that changes, so to assume a fixed value would not account for the changes in pressure and magnetic flux (for tokamaks) that would accompany changes in ion temperature and certainly affect the nature of the time dependence of the burn processes.

In the context of fusion burn, consider the fact that changes in particle and energy density would alter the occurring reactions and thus, the injection rates (adding heat by means of resistive heating, radio-frequency heating, and neutral beam injection, among others), resulting variations within the way in which a confined plasma is organised. Under the assumption of a uniformly distributed ion density in a fixed point in space- where ions equally populate all regions of confinement, the dependance of the fuel ions on instantaneous time, or the rate equations, can be given, with both the ion injection and leakage equal to zero, by:

$$dN_d / dt = -R_{dt}(t) = N_d(t)N_t(t) <\sigma v_{>dt}$$

and

$$dN_t / dt = -N_d(t)N_t(t) <\sigma v_{>dt}$$

Assuming a 50:50 split of deuterium and tritium, the rate equations can be combined to give the total fuel ion density at an arbitrary time during the burn as

$$dN_i / dt = - \{[<\sigma v_{>dt} N_i^2(t)] / 2\}$$

The burn times remain short for inertial confinement reactors (of the order of $10^{-8}$s) and relatively long for magnetic confinement systems (of the order of one second or more). In addition, it is possible to denote the isothermal fuel ion density variation with time with the following relation:

$$N_i(t) = 1 / [(1 / N_{io}) + (½) <\sigma v_{>dt} t]$$

If it is the case that the temperature of the ions is constant during the burn time ($<\sigma v_{>dt}$ = constant), then bremsstrahlung radiation or external heating must be involved. The fraction of fuel burnup in the plasma is bound to be given by the following, and share a relationship with the temperature of the ions as follows:

$$F_b = [N_{i,0} - N_i (\tau_b)] / N_{i,0} = \{1 / 1 + [(N_{i,0} / 2) <\sigma v_{>dt} \tau_b]\} -1$$

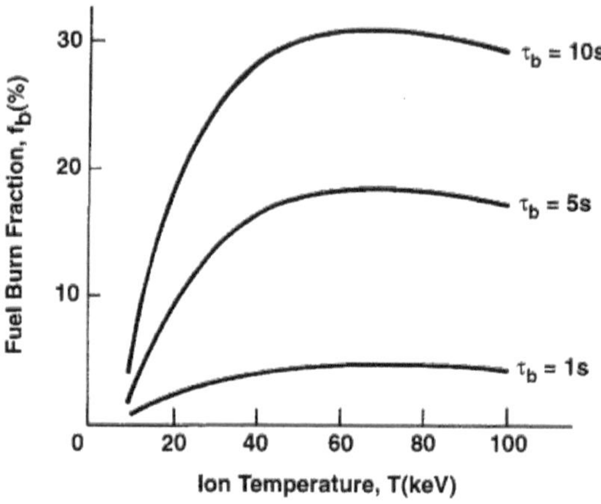


Figure 5: Fusion fuel burn fraction as a function of ion temperature

Furthermore, the fusion power released in a fixed unit volume, as well as its relationship with confinement time as compared to the fuel ion density are given below:

$$P_{dt}(t) = [N_i^2(t) / 4] <\sigma v_{>dt} Q_{dt} = \frac{1}{4} \{1 / [(1/N_{i,0}) + (\frac{1}{2}) <\sigma v_{>dt}(t)]\}^2 <\sigma v_{>dt} P_{dt}$$

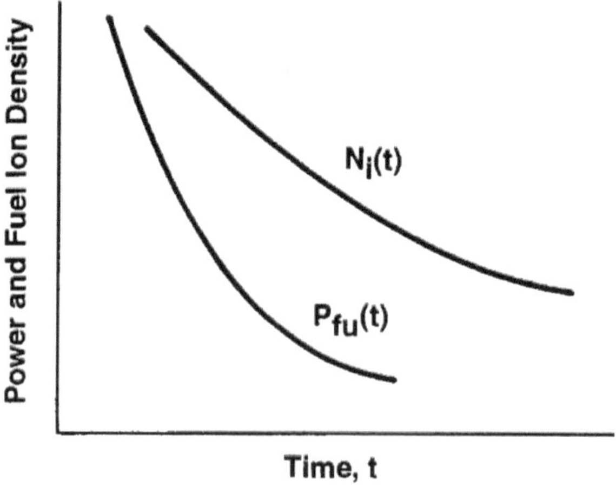


Figure 6: Dominant variation of the power and fuel ion density with time

Now for a more comprehensive analysis of D-T burn. Such analysis must comprise the fusion reactions themselves, along with the injected and leakage rated from the unit volume of interest. Current research aims to determine the time quantity $\tau$ based on other parameters, however, for the current purposes $\tau_j$, denoting the characteristic time parameter representing the local mean time of the residence of the ions in the plasma, is taken as a known parameter. Evidently, the global particle confinement time is equal to the average density-weighted volume of the local mean residence time, making more complete fuel balance equations giving as the following couples set

$$dN_d / dt = F_d - N_d(t) N_t(t) <\sigma v_{>dt} - [N_d(t) / \tau_d]$$

and

$$dN_d / dt = F_t - N_d(t) N_t(t) <\sigma v_{>dt} - [N_d(t) / \tau_t]$$

The combination of these two yields what's known as Ricatti's equation- a dynamic equation for the density of the fuel ions at a fixed point in space and some time that can be written, with $a_0=F_i$, $a_1=-1/\tau_i$, and $a_2=-<\sigma v_{>dt}/2$, as

$$dN / dt = a_0 + a_n N +_{a2} N^2$$

Thus, when the parameters are constant, the equation can be rearranged and subsequently utilised to determine the favorability of a given fuel ion density, as follows

$$N_i^0 = [(-1/\tau_i) + \sqrt{[(1/\tau_i^2) + 2 F_i <\sigma v_{>dt}]}$$

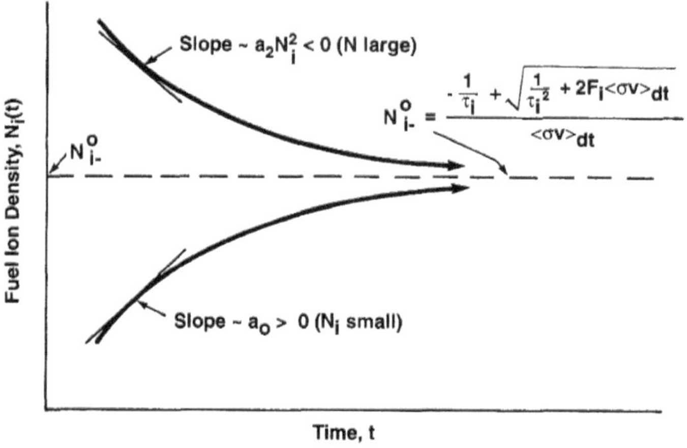


Figure 7: Representation of the fuel ion density as a function of time

## Identical Particle Burn

Considering how a deuterium-tritium fuel combination is highly favourable due to its large reaction cross section, high energy release, and the fact that it requires comparatively lower temperatures to fuse, it is likely to be used for the first generation of fusion reactors. However, without the possibility of breeding tritium in the walls of a fusion reactor or in the control rods of a fission reactor, it is unlikely that future reactors would be able to employ it. And if the mentioned methods are able to meet the current demand for tritium in fusion research, the need for them will inevitably limit the proliferation of fusion technology.

Hence, it would be suitable to consider other options such as d-d fuel which would be able to avoid all the problems associated with radioactive tritium breeding. However, as mentioned before, such a reaction can yield tritium as a product, so the need to handle it is not absent. Moreover, the tritium product can react with the original deuterium fuel, thus resulting in the standard reaction considered above as well as those which pertain to d-d reactions (others such as d-p and p-p reactions require much more extreme conditions, making the probability of their occurrence close to zero).

Starting with the general characteristics associated with using identical particles for fusion reactions. In this case, a = b, making the total distinct reaction possibilities:

$$N_i^2 - N_i / 2 = ½ N_i (N_i - 1) \approx N_i^2/2$$

Here it is assumed that the case is one of general interest, and hence, $N_i >> 1$. With the relation above, it is possible to frame the equation for the reaction rate density by adding the term $N_a^2$, which refers to the total number of indistinguishable ions per unit volume, as such:

$$R_{aa} = < \sigma v_{>dd} (N_a^2/2)$$

Evidently, then, $R_a$ differs from its two fuel counterpart by a factor of two. A severe limitation that characterises D-D burn modes is the fact that for temperatures up to 200 keV, the reaction parameter Sigma-V is much lower than it is for its D-T. In fact, the parameter is several orders of magnitude lower than its counterpart. This means that either a D-D reactor will have to increase the volume of its plasma in such a way that it can compensate for its lower power density. However, this would be associated with its own challenges, ranging from the need for a larger reactor to the formation of instabilities, thus, making the added expense better spent on D-T reactors- whose size can also be increased in much the same way to improve confinement.

Another option is to achieve conditions above 200 keV in the reactor, putting D-D' power density on par with that of D-T. However, such temperatures are associated with radiation losses, and hence, don't comprise the optimal circumstances for a fusion plasma. Here it is useful to consider the reaction channels of a D–D reactor. As mentioned previously, there are two possible reactions, both of which are almost equally likely to occur:

$$\mathbf{d + d \rightarrow t + p}$$
$$\mathbf{d + d \rightarrow {}^{3}He + n}$$

However, there are other dominant reactions that occur such as

$$\mathbf{d + t \rightarrow a + n}$$
$$\mathbf{d + {}^{3}He \rightarrow a + p}$$

As alluded to earlier, a useful parameter for determining the fusion reaction rate with a given set of fuels is called sigma-v. It consists of the cross-sectional area of a particular fusion reaction (σ) multiplied by the relative velocity (v) of both the colliding particles. As such, higher values of σv indicate that a certain reaction is more favourable, as each collision between the particles is more likely to result in a fusion reaction. The σv parameter is typically expressed in units of barns-centimetres per second (barn·cm/s) or metres squared per second ($m^2/s$), where 1 barn is equal to 10^-28 square metres. Such a parameter is highly useful for comparing fuels, and as such, will be utilised later for that very purpose.

A sigma-v ion temperature graph with all the D-D reaction channels evidently indicates the propensity for D-T reactions to occur at a far more favourable rate, followed by D-D and others such as D-H, $^3$He-N, and $^3$He-$^3$He in that order (see fig. 8) Note however, that the very reason for the existence of the sigma-v parameter is to allow for reactors to be optimised for the fuel they are expected to use. For instance, D-T reactors don't heat the plasma beyond a certain threshold determined by the value of the σv parameter to reduce the need for the consequent stronger magnetic field needed for maintaining confinement. Moreovcr, in such cases, reactors are designed to primarily employ ohmic heating as compared to neutral beam injection so as to utilise the higher pressure. However, such design choices are harder to make for a D-D reactor by virtue of the change that occurs over time in the concentration of deuterium, titanium, helium-3, and hydrogen (proton).

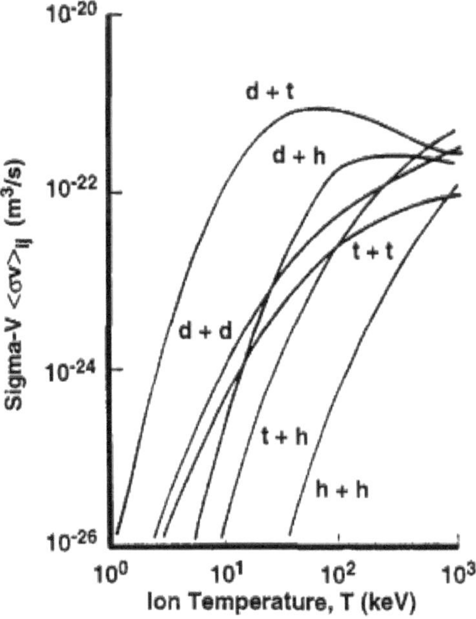


Figure 8: Sigma-v for various fuels defined by a Maxwellian ion distribution

On that note, it is also important to account for less likely through still possible reactions between the products. These include T-T, $^3He$-$^3He$ (mentioned earlier), and T-$^3He$, with the first two having a single reaction channel and the last one two as such:

$$t + t \rightarrow 2n + a$$

$$^3He + {}^3He \rightarrow 2p + a$$

$$t + {}^3He \rightarrow p + n + a$$

$$t + {}^3He \rightarrow d + a$$

A deuterium-deuterium sustained reaction system can be classified into 4 distinct modes. These are a Pure-deuterium mode, a semi-catalysed deuterium cycle, and a catalysed deuterium cycle.

Pure-D Mode: In this mode, the reaction rates for the two possible channels can be given respectively by

$$D + d \rightarrow t + p$$

$$R_{dd,t} = (N^2d / 2) <\sigma v_{>dd,t}, Q_{ddt} = 4.1 \text{ MeV}$$

and

$$D + d \rightarrow {}^3He + n$$

$$R_{dd,3\text{-}He} = (N^2d / 2) <\sigma v_{>dd,3\text{-}He}, Q_{dd,3\text{-}He} = 3.2 \text{ MeV}$$

Note that here and later in instances where helium-3 is written as a subscript, it will be denoted by 3-He. The values for sigma-v used above are provided here at a temperature of common interest:

$$<\sigma v_{>dd} = <\sigma v_{>dd,t} + <\sigma v_{>dd,3\text{-}He}$$

$$<\sigma v_{>dd,t} \approx <\sigma v_{>dd,3\text{-}He} \approx \frac{1}{2} <\sigma v_{>dd}$$

SCAT-D Mode: Here the tritium that is bred as a result of one of the D-D reaction channels will be quickly consumed in a D-T reaction due to the fact that the $<\sigma v_{>dh}$ parameter is significantly larger as compared to $<\sigma v_{>dt}$. Thus, with the assumption that $^3He$ will not react with any deuterium, the initial D-D channel that yields tritium and the subsequent D-T reaction can be summarised as follows:

$$5d \rightarrow 2n + {}^3He + a + p, \; Q_{SCAT\text{-}D} = 24.9 \text{ MeV}$$

If the assumption that both of these reactions occur at equal rates is made, ie. $R_{dd,t} = R_{dt}$, then the following approximation can be framed:

$$N_t / N_d = \tfrac{1}{2} (<\sigma v_{>dd,t} / <\sigma v_{>dt}) \approx \tfrac{1}{4} (<\sigma v_{>dt} / <\sigma v_{>dt})$$

Overall what is to be taken away is the fact that at low-to-medium temperatures (in the context of fusion reactors), the system will operate in SCAT-D mode, yielding reaction products as if two reactions are principally occurring simultaneously. The larger number of deuterons would seem to imply, at least initially, that D-D reactions would occur more frequently as compared to D-T. However, the frequency at which the latter occurs compensates for this, making the assumptions upon which the above approximation is predicated satisfactory for the purposes of this book.

CAT-D Mode: Passing a certain threshold of ion temperature, a fusion reactor becomes more likely to fuse deuterium and helium-3 as compared to tritium. What results is known as a complete catalysing process, as the tritium product fuses with the deuterium fuel, while at the same time, the $^3He$ product of the other reaction channel does so as well. This can be summarised, with the assumption that all 4 reactions are involved at equal rates, as follows:

$$6d \rightarrow 2n + 2a + 2p,\ Q_{CAT\text{-}D} = 43.2\ \text{MeV}$$

It should be noted that every D-D burn mode is challenging to sustain individually. The high temperatures common to all make them prone to needing extremely strong magnetic fields. Moreover, the propensity for ions to jump across field lines increases with average kinetic energy- something which requires a larger reactor to sustain during the burn time. However, changing between the burn modes here can offer a solution to the problem by allowing for different fuel combinations to support ignition as ion temperature continues to increase.

Any further comment on this would be delving into the realm of speculation, as no large-scale reactor capable of sustaining a burn time that allows for the transition of the fusion plasma from the pure to SCAT to CAT modes currently exists or is planned for the near future. Such a scheme has been proposed, however, and its reaction channel would appear, with particle energy (MeV) in brackets, along these lines:

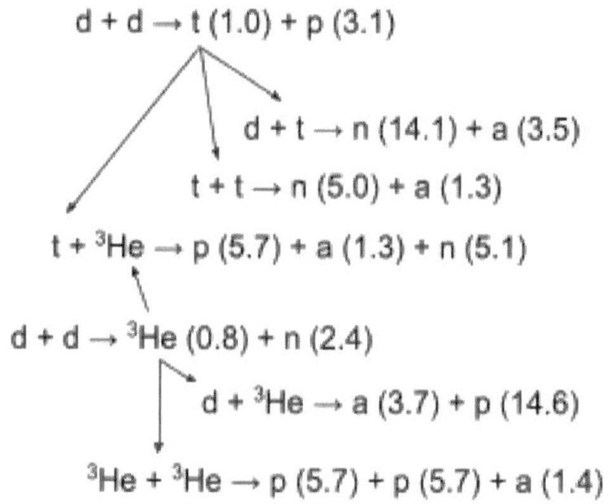

$d + d \rightarrow t\ (1.0) + p\ (3.1)$

$d + t \rightarrow n\ (14.1) + a\ (3.5)$

$t + t \rightarrow n\ (5.0) + a\ (1.3)$

$t + {}^3He \rightarrow p\ (5.7) + a\ (1.3) + n\ (5.1)$

$d + d \rightarrow {}^3He\ (0.8) + n\ (2.4)$

$d + {}^3He \rightarrow a\ (3.7) + p\ (14.6)$

${}^3He + {}^3He \rightarrow p\ (5.7) + p\ (5.7) + a\ (1.4)$

As the CAT-D mode indicates the possibility of deuterium-helium-3 fusion beyond the temperature range where the high sigma-v parameter of D-T makes it the dominant reaction, the viability of this reaction will be considered. To do so is necessary considering the problems intrinsic to D-T fusion. Firstly, the production of radioactive tritium raises safety concerns regarding the process of replacing the reactor walls-something which is inevitable due to neutron collisions creating enough holes for the vacuum to be compromised. This increases the costs associated with fusion energy as a source of energy, and if fusion is to compete with fossil fuels and renewables it must produce extremely large amounts of energy or find alternate fuel sources. Whether there is a limit to energy production, or aneutronic reactions are in some fundamental way more viable for fusion is yet to be seen, so making a conclusive choice between them is currently impossible.

## D-$^3$He Fusion

Evidently, D-D reactions, the next best option, are not truly aneutronic, as not only does one reaction channel yield a neutron as an end product, but in all possible burn modes some part of the end product consists of neutrons- a single neutron in the pure-D mode and 2 for both the SCAT-D and CAT-D modes. Thus, there are only two truly aneutronic reactions that possess a sigma-v parameter suitable for power generation purposes- $^3$He-$^3$He and d-$^3$He. The latter, due to its better performance in that regard, will here be considered:

$$\mathbf{D + {}^3He \rightarrow Q_{d3\text{-}He} = 18.3\ MeV}$$

Since $<\sigma v_{d3\text{-}He}$ is at a higher temperature as compared to the maximum value of $<\sigma v_{dt}$, the d-$^3$He reaction requires a significantly higher temperature in order to occur. Another consideration is the fact that the emission of bremsstrahlung radiation ("braking radiation"- radiation given off by free electrons that are deflected (i.e., accelerated) in the electric fields of charged particles and the nuclei of atoms) is more severe due to the higher number of protons in helium as compared to hydrogen. Finally, the problem of the shortage of tritium cannot be solved with this fuel, as helium-3 is also rare, having a natural abundance of $^3He/(^3He + ^4He) \approx 10^{-6}$. There are, however, methods by which helium-3 can be created or sourced. Firstly, tritium, by a process known as nuclear decay- the emission of energy in the form of ionising radiation, becomes helium-3 as follows:

$$t \rightarrow {}^3He + \beta$$

This method cannot be scaled, however, as tritium has a half-life of 12.3 years. As it happens, this value is too short for tritium to be reliably stored, while at the same time, it is too long for it to be stockpiled as well. Further, as seen above, $^3He$ can also be the product of D-D reactions. Finally, there is the possibility of sourcing large quantities of it from the moon. This is supported by the fact that $^3He$ has been identified in lunar rock samples. Under energetically favourable conditions, it can be mined on the moon and transported to Earth.

However, that deuterium can react with itself, along with the fact that it can do so at lower temperatures than with helium-3, makes the D-$^3He$ reaction not fully "clean". The existence of $N_d$ (a certain number of deuterium particles) allows for neutron and tritium-producing reactions, with their reaction rates being as follows

$$d + d \rightarrow t + p, \; R_{dd,t} = <\sigma v_{>dd,t} (N^2_d / 2)$$

$$d + d \rightarrow {}^3He + n, \; R_{dd,h} = <\sigma v_{>dd,n} (N^2_d / 2)$$

Ultimately, whether this makes D-$^3$He fuel no better than D-T boils down to the ratio of clean reactions relative to unclean ones. Such a ratio can be expressed for the tritium-yielding as well as $^3$He yielding reaction channels respectively as

$$R_{dh} / R_{dd,t} = 2 (<\sigma v_{>dh}/<\sigma v_{>dd,t}) (N_h/N_d)$$

and

$$R_{dh} / R_{dd,3\text{-}He} = 2 (<\sigma v_{>dh}/<\sigma v_{>dd,3\text{-}He}) (N_h/N_d)$$

These equations indicate the importance of temperature and the ratio of $^3$He to deuterium in an ion population. Firstly, the reaction rates of the neutronic and aneutronic reactions vary, providing temperature ranges wherein one type of reaction is more likely than the other. D-D reactions are able to occur at lower temperatures due to their high sigma-v parameter, thus making them more likely at lower ion temperatures. Moreover, what is of equal importance is the split of the two ions. Having $^3$He account for a greater percentage of the total ion population would evidently make its reaction with deuterium more likely than the reaction of the latter with itself. However, depending on the reactor conditions, this is likely to limit the reactions possible, as beyond a point there will be no more deuterium for the remaining $^3$He ions to fuse with- thus limiting the number of maximum reactions possible to the number of deuterons. Moreover, neither do the D-$^3$He or $^3$He-$^3$He reactions produce viable fusion reactants- particles which can further fuse with $^3$He or even deuterium. If a reactor is to operate in a "clean" manner, then its fusion cycle must have good control over the temperatures achieved as well as the composition of the fuel ions.

## Spin-Polarised Fusion

Another aspect of the motion of an ion is its quantum properties, those which affect its motion in a manner not accounted for by their interpretation in terms of collective behaviour. One such example is the spin of a particle, a property intrinsic to particles such as protons, neutrons, and electrons, that is somewhat analogous to the angular momentum of a rotating object. However, it is not the result of actual physical motion but rather a quantized (in discrete values) fundamental property that has no classical counterpart.

In the context of fusion, along with the considerations to do with collisions and an increasing reaction rate, there is also the modification of the arrangement of particles in a system based on the alignment of their spins in the same direction. In this case, the particles are said to be polarised. This phenomenon is currently being explored as a means by which to suppress less-desired side reactions, and thus, improve overall efficiency. How this occurs under the conditions of interest will now be explored.

For neutrons and protons, spin is characterised as ½ $\hbar$, where $\hbar$ is Planck's constant. The vectorial addition of the spin of nucleons which form nuclei would show that even a nucleus, then, has a spin. The spin of nucleotides involved in the d-t fusion process is as follows (in units of Planck's constant $\hbar$):

Deuteron: 1

Triton: ½

Neutron: ½

Alpha: 0

Here the direction of spin is measured relative to the external magnetic field. It would be possible to denote the spins for deuterons and tritons with + representing the direction parallel to the magnetic field, - the direction anti-parallel, and 0 for the transverse direction, along with $f^0_0$ for the fraction of nuclei oriented in directions permitted by quantum mechanics as follows:

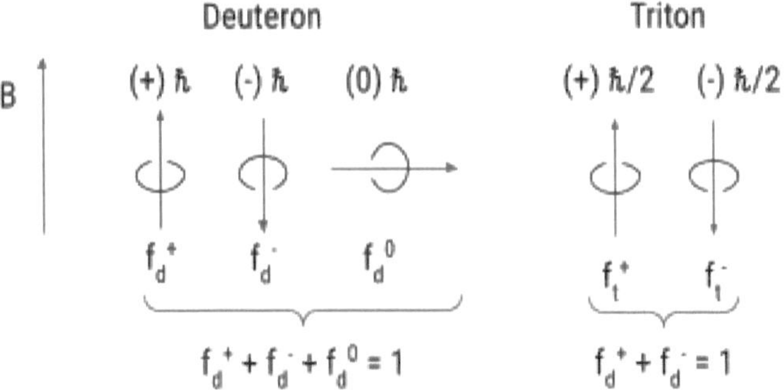


Figure 9: Spins for Deuterons and Tritons

Nuclear transformations, on the whole, are characterised by the conservation of spin. The maximum spin for $^3He$ is 3/2 (in units of ħ) and occurs when the spin of the colliding deuteron and triton align when the reaction is:

$$d + t \rightarrow n + a$$

It would be possible to express the cross section with fractional concentrations of deuterium and tritium in the spin states which are allowed by quantum mechanics by using the notation from above such that

$$\sigma_{dt} = [(f_d^+ f_t^+ + f_d^- f_t^-) + ⅔ (f_d^0) + ⅓ (f_d^+ f_t^+ + f_d^- f_t^-)]\ \sigma_0$$

With this equation is it possible to evaluate how different mixes of polarisation of deuterium and tritium in a magnetic field would affect the reaction cross section. More specifically, it would be possible to notice under what conditions $\sigma_{dt}$ approaches the maximum fusion cross section possible $\sigma_0$.

Interestingly, taking an equal fraction of fuel nuclei in their allowed spin states results in a completely depolarised state. The ratios which define such a state, as well as the reaction cross section $(\sigma_{dt})_{r,}$ of two particles (deuteron and triton) with said given ratios can be respectively denoted by

$$f_d^+ = ⅓ , f_d^- = ⅓ , f_d^0 = ½ , f_t^+ = ½ , f_t^- = ½$$

and

$$(\sigma_{dt})_{r,r} = \{[(⅓) (½) + (⅓) (½)] + ⅔ (⅓) (½) + (⅓) (½)]\} \sigma_0 = ⅔ \sigma_0$$

Here note that r is not only used to denote random but placed twice to account for both fuels exhibiting such polarisation states. Later both fuels will experience different spin polarisation, for which such a form of notation is useful.

If a random spin polarisation is assumed, ie. If no attempt is made to regulate this property of the fusion nuclei, then the cross-section $\sigma_{dt}$ will be equal to two-thirds of $\sigma_0$. However, the effect of spin polarisation can vary based on the conditions under which nuclei are fusing. For instance, for a parallel arrangement of deuterium and tritium nuclei, the fractional concentrations of deuterium and tritium in their allowed spin states, and subsequent cross-section $(\sigma_{dt})_{+,+}$, are evidently

$$f_d^+ = 1\ ,\ f_d^- = 0\ ,\ f_d^0 = 0,\ f_t^+ = 1\ ,\ f_t^- = 0$$

and

$$(\sigma_{dt})_{+,+} = (1 + 0 + 0 + 0 + 0)\ \sigma_0 = \sigma_0$$

This clearly shows how a fifty per cent increase in fusion reaction rate density can be achieved with deuterium-tritium fuel by having all deuterons and tritons having a spin alignment in the direction of the magnetic field. However, the same outcome can be achieved if all spins are in the opposite direction. This is, of course, an ideal circumstance that would be difficult to create in the fusion plasma, where the nuclei behave in accord with the state of random spin polarisation described above.

The closest means of exploiting the spin property of fusion nuclei, then, at least with current fusion devices, is through the injection of nuclei with specific spin polarisation. For most reactors with a longer burn time, such as those which employ magnetic confinement, a chamber of tritium with nuclei of random spin injected with deuterons of a specific spin polarisation (in the direction of the magnetic field) appears to be a practicable method by which the phenomena described above can best be exploited, and hence will now be considered. Evidently, within such a circumstance the fraction of nuclei oriented in allowed spin states and the expression of cross-section $\sigma_{dt}$ in terms of said fractions are given respectively as

$$f_d^+ = 1\ ,\ f_d^- = 0\ ,\ f_d^0 = 0,\ f_t^+ = ½\ ,\ f_t^- = ½$$

and

$$(\sigma_{dt})_{+,r} = (½ + 0 + 0 + ⅓\ (½) + 0)\ \sigma_0 = ⅔\ \sigma_0$$

Accelerators capable of supplying spin-polarised deuterium or tritium ions into a fusion plasma do indeed exist, increasing the prospects of such a scheme into the realm of practicality. Moreover, even with roughly half the ions in a spin-polarised state, fusion power density can be increased by fifty per cent as such a state can be sustained for up to ten seconds.

However, for a reactor to be able to facilitate polarised fusion, of this kind or any other, special considerations need to be made for the containment wall. This is due to the preferred directional distribution the neutron and alpha particle reaction products assume under such conditions that result in select spots of the chamber being subject to disproportionate neutron bombardment. Moreover, the alpha particle actively carries some of the energy in a D-T reaction, and hence, under such conditions heats up select portions of the plasma.

In closing, spin-polarised fusion is certainly a viable method for increasing the fusion power density in a reactor. There are, however, considerations that require any reactor making use of it to be designed specifically to do so. At this stage it is impossible to comment on whether the injection of ions of a specific spin polarisation can yield a large enough improvement to the rate of reactions is justified in context to the overall energetics of the reactor. Not to mention, the complete implications of re-designing a reactor's structure to accommodate directional distributions of the kind mentioned above are unknown, and the possibility that to do so would undermine the underlying principles of its design is very real. Finally, the analysis above was limited to a fuel consisting of deuterium and tritium, which, while appropriate under normal circumstances, is hardly apt for a phenomenon so distinct from others which afflict the constitution of a fusion plasma; it could very well be the case that the propensity of alternate fusion fuels, otherwise not considered due to their low reaction cross sections within temperatures of interest, to exhibit certain spin states would make them respond to spin-polarised ion injection better- perhaps even to a point of greater viability- than D-T fuel.

## Catalysed Fusion

So far all but the recent section have described states of operation that are not only known to be feasible but have been and are being considered all over the world as practicable methods to achieve a fusion energy gain factor greater than zero. Moreover, all concepts described hitherto pertain to some degree to all, or at least the majority of fusion reactor concepts. However, the concept of catalysed fusion is mainly considered for so-called cold fusion- a hypothesised type of nuclear reaction that would occur at, or near, room temperature. Until recently the scientific community has been averse to any claims of cold fusion making progress, although the use of a muon as a catalysing agent has increased its prospects. The means by which a reactor can employ muons to facilitate low-temperature reactions (those occurring at temperatures under or near 300°C), is one of numerous recent cold-fusion concepts. Here, the general principles of the use of a catalysing agent to increase the reaction rate for deuterium-tritium fuel, with the equations, however, being applicable for any combination of fuel ions, will be considered.

It is first important to understand the role of a catalysing agent in the fusion process, for which it is useful to return to the reaction statement of the conversion of mass to energy from Chapter 1

$$E_{dt} + d + t \rightarrow (dt)^* \rightarrow n + a + Q_{dt}$$

Here (dt) refers to the intermediate state $^5He$ that decays almost immediately into the neutron and $^4He$ reaction products, while, as before, $E_{dt}$ refers to the energy supplied to heat the D-T fuel to thermonuclear temperatures as well as sustain such kinetic energy against loss mechanisms so that particle collisions have enough force to overcome the coulomb barrier. Evidently, a condition for viability is $E_{dt} < Q_{dt}$. However, the need for extreme temperatures in achieving fusion that accounts for a large $E_{dt}$ value can be overcome with the use of a catalysing agent added to the plasma to neutralise coulomb repulsion. This would then enable ions to enter the range of the strong force.

A catalyst in this sense is analogous to a chemical catalyst- a compound which enhances the rate at which a chemical reaction occurs. Such an agent must be released after a collision event and then proceed to induce another reaction. It would be useful to incorporate such an agent into the above equation. If the energy required to produce 1 agent x is $E_x$ and a single particle of this agent (1x) catalyses X fusion events, the sequence of energy transformation is:

$$E_x + d + t + x \rightarrow (xdt)^* \rightarrow x + n + a + Q_{dt}$$

$$d + t + x \rightarrow (xdt)^* \rightarrow x + n + a + Q_{dt}$$

$$d + t + x \rightarrow (xdt)^* \rightarrow x + n + a + Q_{dt}$$

$$\vdots \qquad \qquad \vdots \qquad \qquad \vdots \qquad \vdots$$

$$X\,Q_{dt}$$

Based on this the requirement for energy viability is $E_x < XQ_{dt}$. A muon ($\mu$) can be considered as a catalyst option, as not only does it have the properties suggested above, but it is also the product of various types of high-energy nuclear reactions. Before considering the use of a muonic molecule to facilitate fusion reactions at lower temperatures, it is first important to summarise the properties of a muon which determine its behaviour. Firstly, a muon's charge ($q_\mu$) is the same as an electron, which in terms of magnitude is approximately $-1.6 \times 10^{-19}$ Coulombs. However, a muon's mass $m_\mu$ is significantly greater than that of an electron- approximately 207 times greater($\approx$105.7 MeV/$c^2$). Moreover, it is highly unstable, with a lifetime $\tau_\mu$ of only around $2.2 \times 10^{-6}$ s. As a result of its greater mass, it is able to enter 207 times as small a Bohr radius of a hydrogen atom as compared to an electron (see fig. 10). This occurs due to the fact that while electrons and muons experience the same amount of repulsion due to their identical charges, the larger mass of the latter makes it harder to repel.

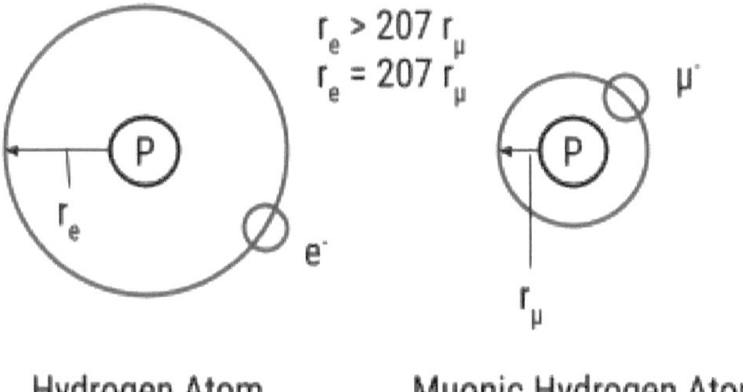


Figure 10: Muonic hydrogen with a Bohr radius 207 times smaller

A consequence of this is the fact that a muonic hydrogen atom appears as an oversized neutron. Thus, it is able to approach another muonic atom with much closer proximity due to the reduced repulsion. Accordingly, it is possible to bring two muonic atoms close enough for the strong force to take effect and incur a fusion event.

Now that the principle behind the use of muons and the formation of muonic atoms is established, the process by which muons can act as a catalysing agent for deuterium-tritium fusion will now be considered. Figure 11 provides a schematic of a muon-catalysed D-T reaction sequence:

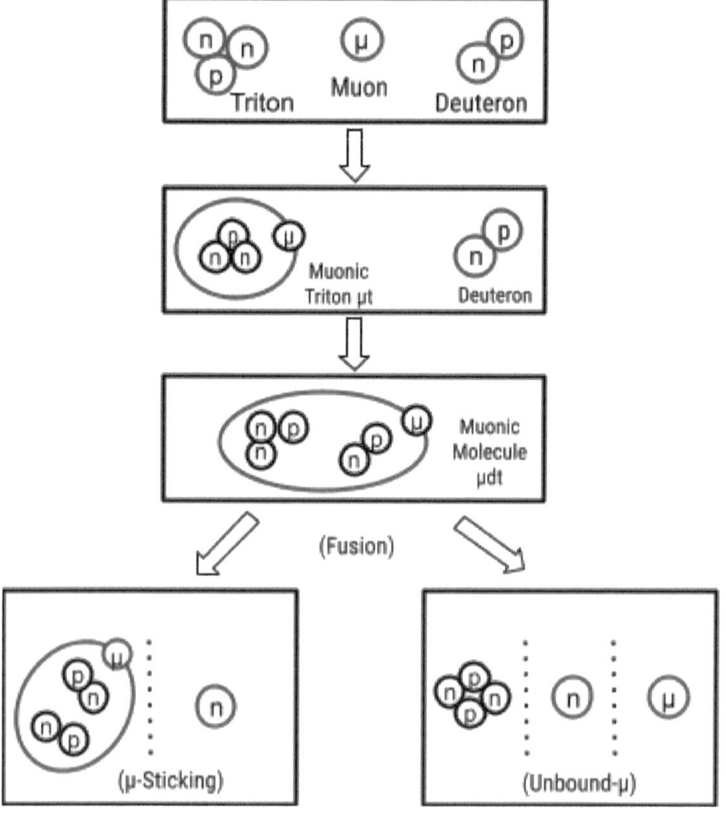


Figure 11: Muon catalysed D-T reaction sequence

Fusion reactors of the sort depicted above have been experimentally confirmed in liquid hydrogen between 300-900 Kelvin. Furthermore, such a reactor could be sustained in a fission reactor environment. And as far as an independent muon-fusion reactor is concerned, the configuration of its chamber and muon-producing accelerator would be similar to that of an inertial confinement scheme sustained by an ion beam. However, the short lifetime of muons constitutes a major limitation of such a method of fusion, as it remains whether the muon is bound to a nucleus or not, and hence, requires a fast reaction rate during the entire course of the burn time.

Overall, catalysed fusion is, like any emerging, not without a unique combination of strengths and limitations. However, if this section has established anything it is that the prospects of thermonuclear fusion at lower temperatures haven't disappeared, and it is very much a possibility that fusion reactors become more energy efficient not through higher gains, but by lower expenditures in terms of the energy needed to cause the initial fusion of nuclei.

# Fusion Energetics

A fusion energy system contains and transforms energy in and into different forms respectively. The understanding of the magnitudes of these different forms, as well as how changes in each energy component occur, is a cornerstone of the development of viable reactors. Moreover, there are various methods of heating a plasma to fusion conditions, with the ultimate goal of having the reactions sustain themselves. The energy transformations involved in such a process require an extensive review, for there exist numerous ideas on how to bring about such a condition in the plasma, while practical reactors will likely employ only a few. Thus, it is necessary to consider the energy transformations on a fundamental level, defining what constitutes a desirable energy system while outlining the steps needed for a reactor to achieve a so-called balanced state. Hence, this section will principally consider the underlying conditions for breakeven and ignition. It will then proceed into a discussion of the various methods by which the methods of supplying energy to the reactor can be optimised in order to achieve said conditions. Finally, comprehensive principles will be derived from an analysis of the behaviour of fusion plasmas in response to changes in temperature and pressure as a result of external as well as internal heating methods- shedding light on how a fusion reaction chain can be sustained.

## System Energy Balance

In the discussion of energetics it is first essential to consider what constitutes a balanced system in terms of the energy it consumes and releases. A fundamental requirement for fusion reactors is that during time interval τ the total energy recovered $E_{out}$ must be greater than that supplied/provided $E_{in}$:

$$E_{net} = E_{out} - E_{in} > 0$$

To simplify, the unit of electrical energy can be used for thermal, kinetic, radiation, and magnetic energy. Later when considering the Lawson criterion the energy transformations that occur within a general fusion power plant will be discussed, but for current purposes, the following diagram can depict its associated energy components:

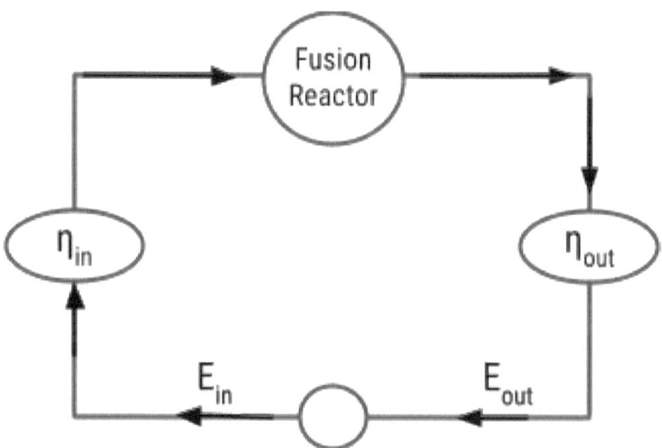


Figure 12: Energy components of a General Fusion Power Plant

The energy actually delivered to the reaction chamber differs from the energy supplied $E_{in}$ by a factor of $\eta_{in}$. Similarly, the energy recovered by $E_{out}$ differs from that collected by the reactor walls by a factor of $\eta_{out}$, although for convenience the $\eta_{out}$ can be a number greater than 1 so that the relationship between $E_{out}$ and $\eta_{out}$ can be described in the same way as $E_{in}$ and $\eta_{in}$- the energy released from the fusion domain differing from $E_{out}$ by a factor of $\eta_{out}$. Thus, η is a term added to account for the inefficiencies associated with the energy transformation processes, and its inclusion depicts the efficiency of converting different forms of energy as well as of coupling specific forms of energy into the plasma. While $\eta_{in}$ is determined by the interaction of an external heating method with the plasma, $\eta_{out}$ in most cases is known to be typical for the employed energy conversion cycles; the system by which energy is here transformed is not affected by the behaviour of the plasma and hence remains both well-known and irrespective of alterations in design (unless a design choice for some reason prohibits the function of the wall). Fusion reactors require a point to be reached in the burn cycle where the average kinetic energy of fuel ions can be increased by collisions with reaction products in order to reduce the amount of energy required to sustain fusion conditions. Moreover, there exists a certain energy threshold particles must possess in order to get close enough to each other to fuse. Hence, there exists a critical plasma energy concentration which fusion reactors must achieve, both for breakeven and ignition. This is unlike fission reactors, which only require a critical mass by being able to sustain a fission chain through particle feedback in the form of released neutrons. On the other hand, the stages necessary to maintain a fusion chain by energy feedback are as follows:

Current reactors, however, fail to achieve the critical energy concentration in the plasma, at least not in a manner that is sustainable. Various heating methods will be considered in the subsequent section, but in the context of this discussion, the distinction will be made between external (auxiliary) and internal (fusion power deposition) methods. The role of these methods is to combat losses that occur as a result of radiation and other forms of leakage. Thus, accounting for the efficiency of the input energy coupling to the plasma ($\eta_{in}$), the energy of auxiliary heating can be expressed as:

$$E_{aux} = \eta_{in} E_{in}$$

In the fusion cycle considered, the following expression and its rearrangement, with $f_c$ denoting the fraction of charged particles in the fusion product energy, remains constant:

$$E_n / E_{fu} = 1 - f_c$$
$$E_{fu} - E_n = f_c E_{fu}$$

Here it is possible to include the fundamental parameter used to measure the efficiency of the convention of the energy deposited into the plasma into fusion energy. This criterion, sometimes called the plasma fusion multiplication, but chiefly referred to as the Q value, can be expressed with the above terms as:

$$Q_p = E_{fu} / \eta_{in} E_{in}$$
$$= E_{fu} / E_{aux}$$

Evidently the Q value signifies whether the reactor simply breaks even or produces a net energy gain. If the value of Q is 1, then the power being released by the fusion reactions is equal to the required heating power, whereas if it is greater, then, so long as the amount by which it is greater is not less than the energy lost in the reactor walls, it is greater. The specific nature of the interaction between the density of a plasma and its temperature is complex, and the methods by which to account for it will be later visited. For the case of D-T fusion, assuming ignition is happening everywhere in a homogeneous plasma, with the thermal plasma energy density here is as $E_{th,j} = 3/2\ N_i\ kT_j$ , j=i,e, then the ignition condition can be given by:

$$F_{c,dt}\ P_{dt}\ (N_i,T_i) = P_{br}\ (N_i,\ N_e,\ T_e) + P_{cyc}^{net}\ (N_e,\ T_e) + 3/2\ [(N_i\ kT_j + N_e\ kT_e) / \tau_E$$

## Plasma Heating

The role of temperature in fusion is that it allows for particles to overcome the strong repulsion they experience when approaching each other enough so that they may fuse. The particles then must have sufficient kinetic energy so that fusion reactors occur. The initial neutral gas becomes ionised when heated- that is, they lose electrons as a result of the fact that the kinetic temperature is greater than the ionisation energy potentials. This means that electrons are stripped from the nucleus and are freely moving around the fluid, leaving positively charged ions or cations in the form of the nuclei. This state of matter is called plasma. However, the conditions under which sustained fusion occurs require temperatures beyond what is achieved in the plasma formation heating process, and hence, temperatures must be raised until a reasonable reaction rate $<\sigma v>$ can be achieved. Such conditions usually necessitate values around the 10 keV range.

It is commonly assumed that a state of plasma is necessary for fusion, however, the temperatures involved are of far greater magnitudes, and hence, much more difficult to sustain than a plasma. The attainment of sufficiently high temperature and preservation of such conditions against losses is collectively referred to as a burn cycle. This distinction is made since it places a limit on burn time and outlines the power consumption for confinement.

The first stage comprises the initial heating of the fusion fuels. There are multiple methods to do so, each with its own strengths and limitations. Moreover, the effectiveness and efficiency of the methods varies with factors such as pressure and temperature, so most reactors require multiple methods to support thermonuclear fusion throughout their burn cycle. New methods are under development, but for the purposes of constructing practical reactors in the near future, and thus for the purpose of this book, five of them will be considered: resistive heating, compression, electromagnetic wave heating, beam injection, and internal heating. Each method will be considered in the rough order in which they are typically implemented in the heating mechanisms of most magnetic confinement systems.

### Resistive Heating

Resistive hearing involves the passing of electric current within the plasma. The effect of this is a phenomenon known as ohmic heating. The principle of ohmic energy dissipation effects upon which such a method is based is simple- A loss of electric energy due to the heat generated by the flow of current through a resistance. This is the reason why incandescent bulbs release large amounts of heat. A similar condition is simulated in the plasma, with the plasma acting as the resisting medium. Some calculations of the power that can be deposited will illustrate how this can be achieved. Evidently, in a unit volume of plasma, the power deposited can be given by:

$$P_{res} = \eta\, I^2$$

Here I is the current density while η is the resistivity of the plasma. The latter parameter depends on the number of collisional effects as such:

$$\eta \propto kT^{-3/2}$$

This indicates an inversely proportional relationship between plasma resistivity and temperature- Increases in temperature correspond to reductions in resistivity. This evidently implies that the effectiveness of ohmic heating progressively decreases as the plasma temperature increases. That such is the case has also been experimentally confirmed. The point at which the provision of supplementary heating is typically necessary is approximately beyond 1 keV. It should be noted, however, that not all concepts benefit the same from resistive heating, as fundamental differences in burn time aside, some system concepts allow for the generation of massive currents, and thus have more to gain from such a method.

### Compression

Alluded to earlier as not only a method of bringing fuel ions in the proximity of one another, but for creating the temperature conditions in which they fuse as well, compression refers to the use of either magnetic and/or mechanical forces for the adiabatic compression (no transfer of heat into or out of a system- change in internal energy only done by work) of plasma to raise its temperatures. Compression methods can be categorised into two types: short intervals and long intervals. The terms short and long here are to be understood relative to the speed of the transfer of thermal energy. Based on this, short intervals can be considered those which occur for a period of $10^{-6}$ seconds or less. This type involves complex gas dynamics, as well as shock wave considerations since an implosion is usually involved. This type is the principal heating method in inertial confinement reactors. Next, long interval heating methods usually include some form of adiabatic compression, and thus involve the necessary relation $pv^{\gamma}$ = constant, where γ is the adiabatic gas coefficient. However, these intervals are nevertheless quite short relative to radiation losses that can occur within the fusion plasma.

### Electromagnetic Wave Heating

Electromagnetic wave heating refers to the deposition of energy in the plasma through the use of electromagnetic waves from lasers or radiofrequency generators. For these purposes, plasma is to be treated as the subject of collective interactions exhibiting typical resonance effects. This is so because it consists of an aggregation of numerous moving electrical charges. Thus, the prospect of simply coupling high power (high frequency) waves to the plasma to heat it appears evident. Such electromagnetic coupling via waves with an ion cyclotron frequency $w_{g,j}$ is a favourable application.

Unlike ohmic heating, however, the effectiveness of this method doesn't decrease as the temperature increases, and it has been shown to resonantly enhance ion motion to high enough kinetic energies. In fact, the absorption of electromagnetic energy irradiated increases with higher ion temperature T.

The high energy waves are usually irradiated at frequency $w_{g,j}$, or the harmonics $2w_{g,i}$ or $3w_{gi}$ (30-100 MHz depending on the strength of the magnetic field). Assuming that the frequency used is optimised for the conditions of the reactor to deposit the maximum amount of energy for the least of it used for the production of the waves themselves, then the only room to make this method more efficient is by reducing the distance between the wave antenna and the edge of the plasma. To understand how such a process works it would be useful to consider a tokamak, with the inner solenoid acting as the primary coil and the outer magnets the secondary of a resonant transformer. The plasma, then, is the medium within which electromagnetic waves allow for a voltage to be induced into the secondary from the primary. Hence, in the case of tokamaks, spheromaks, and other similar magnetic confinement concepts, the process of making the method efficient involves reducing the ratio of the inner radius to the outer radius.

However, the resonant frequency for ions in the plasma is different from the one for the electrons. Hence, this heating method can be divided into two types: Ion cyclotron resonance heating (ICRH), and electron cyclotron resonance heating (ECRH). The latter involves frequencies in the range of 28 to 140 GHz. This means that the deposition of electromagnetic waves of the same frequency will lead to a significant difference in the temperatures of the electrons and ions. And while the higher kinetic energy of either type of particle is beneficial, ultimately only the temperature of the ions needs to be raised to obtain sufficient fusion reactivity.

### Beam Injection

Beam injection describes the process of injecting neutral particles or pellets into the plasma to deposit their energy to it by collisional effects. When neutrals collide with particles in the plasma, they transfer their kinetic energy, becoming ionised in the process. Moreover, this impact ionisation occurs through a process known as charge exchange, wherein the neutrals exchange electrons with fuel ions. However, the term collision here doesn't refer exclusively to direct, physical contact, since, as has been previously established, bringing two nuclei close to each other is extremely difficult due to the Coulomb barrier. The energy transfer, then, is facilitated by coulomb collisions- a binary elastic collision between two charged particles interacting through their own electric field.

There are two components which determine the rate at which injected ions transfer their kinetic energy $E_f$ to the plasma. The average rate at which energy is transferred to electrons and ions can be respectively approximated with

$$< (dE/dt) >_{f>i} \approx A_e N_e T_e^{-3/2} E_f$$

and

$$< (dE/dt) >_{f>i} \approx A_i N_i E_f^{-1/2}$$

Here $A_e/A_i$ = constants for the respective species and specified beam particles. Rather conspicuously, ion beams with higher energy mellifluously transfer most of their energy to electrons, while those with lower energy have the propensity to do the same but for thermal ions.

### Internal Heating

Finally, internal heating is the transfer of energy liberated in fusion reactors to the ions and electrons of the plasma. This method was briefly described earlier, and due to its ability to scale with the addition of no external energy, it can lead to a state known as ignition, wherein reactors are able to sustain the conditions which promote them in an endless cycle.

The phenomenon of coulomb scattering mentioned earlier is primarily what allows for particles within the plasma to transfer their kinetic energy to each other. To that end, all the considerations and equations of other methods consider this feature of fusion plasmas a loss- a way for particles to dissipate their energy in such a manner that they are no longer in a state to overcome the coulomb barrier. This would imply that internal heating is fundamentally unviable, and at this stage it is true. However, assuming that the energy dissipated comes from fusion reactions within the reactor and not directly from any of the heating methods mentioned above, then there exists the possibility for the process to have only the losses associated with it in the form of the number of collisions that don't result in fusion. However, the ratio of collisions to near misses can be increased by the deposition of larger amounts of energy within shorter periods of time such that coulomb collisions, on average, involve magnitudes of forces approaching that of the electrostatic repulsion experienced beyond the range of the strong force. Further, consider the possibility of this second set of reactions prompting a third. Under most circumstances, it would seem obvious that the energy would undergo continuous dissipation, say for instance, in the case of an echo. This most certainly would occur if it wasn't for the fact that fusion reactions liberate large amounts of energy. And while it seems that the energy needed to sustain a reaction chain could simply be used to heat the plasma directly, the fact that the kinetic energy needed for nuclei to fuse has a defined amount means that even accounting for the losses due to collisional effects that don't result in fusion, it is possible for these auxiliary reactions to sustain a longer burn time with the energy input requirement decreasing in a geometric progression. Thus, the prospect of this approach is the possibility for it to remove the constraint of producing more power than that consumed by the operation of the reactor itself. Most other forms of power generation have insignificant energy expenditures in order to operate, and the promise of ignition is that it can allow fusion reactors to operate in the same way. Furthermore, even if the energy liberated can only sustain breakeven conditions on the account of losing some energy in collisions that don't result in fusion, done long enough, such a method would allow for the reactor to reach a point where the temperatures achieved facilitate a reaction rate that doesn't suffer as many unsuccessful collisions.

Moreover, the entire effort to produce energy from fusion is predicated on the assumption that the reactions are capable of offsetting the cost of giving rise to them in the first place, and hence, that this is the most optimal heating method. Evidently, such a claim is fallacious on the grounds that many reactor concepts plan to rely on sustained fusion reactions of the sort described here in order to breakeven (making the argument a form of circular reasoning), yet the fact that reactions resulting from internal heating require no additional energy to occur means that they are not to be judged on the basis of a set of reactions resulting directly from external heating methods.

Nevertheless, such a discussion must be grounded in the practical challenges associated with sustaining a state of ignition. The efficiency of energy absorption by the reactor walls is significant in this regard but will be relegated to a more comprehensive discussion later. For now, it is necessary to consider principally the mechanism by which part of the energy will be collected while the rest is used to facilitate the internal heating process. Take the case of deuterium and tritium. Most of the energy is in the neutron, which means that it can easily be collected due to the fact that it has no charge, and hence, is not contained by the magnetic field. The remaining energy is in the alpha particle and hence, stays in the reactor, heating the plasma in a manner akin to that of an injected neutral particle. In fact, the specific energy distribution between the two products of the D-T reaction, along with their properties that allows for one to stay confined and the other to leave, makes it possible for a D-T reactor to achieve a point of break-even earlier in its burn cycle. This is important due to the fact that for internal heating to work, it is essential that energy be deposited in large quantities within short periods of time- something which can be done using external heating methods powered by the energy generated by the reactor. Hence, while internal means of heating can, during the burn time, account for the largest production of energy, they need to be coupled with methods which offer the possibility of a greater increase in energy density within a critical period during the start of the burn cycle.

Such discussion is evidently vague and doesn't account for the challenges associated with a long burn cycle or confinement of a plasma of such hypothetical conditions. Along the same lines, even if the reactor is able to sustain such conditions and the plasma behaves in the manner outlined above, then there is still the challenge of bringing the plasma to the necessary temperature to initiate the ignition process. The exact conditions involved can be fundamentally reduced, aside from temperature, to pressure and confinement time, and the means by which they can all be accounted for, thus, providing some insight into how easy or difficult it might be for a reactor to enter a state of ignition, as well as how advantageous such a state may be if achieved, will now be considered.

## Lawson Criterion

The Lawson criterion provides a test of energy balance for fusion devices. That is, it comprises an algebraic relationship based on the notion that $E_{out} > E_{in}$ during time interval $\tau$- a condition for breakeven established in the first chapter. The value of $\tau$ in this case can be one of two things: For devices that operate in a pulsed mode it is the cycling time while for those with a steady-state operation, it refers simply to the period of time. Before understanding the factors which contribute to energy loss, it is first essential to define what constitutes energy that is recovered. There are forms by which energy can be recovered: Fusion reaction energy release ($E_{fu}$), radiation losses ($E_{rad}$), and the thermal motion of particles ($E_{th}$). Evidently, the addition of these would give the fraction of energy recovered. However, no type of energy conversion takes place with 100% efficiency, and hence, it is necessary to include the term η to represent the respective conversion efficiencies, making the fraction of received energy given by:

$$\mathbf{E_{out} = \eta_{fu}\, E_{fu} + \eta_{rad} + E_{rad} + \eta_{th} + E_{th}}$$

The thermal motion of the particles in this case ($E_{th}$) refers to the energy present in the reactor in the form of the kinetic energy of the particles that comprise the plasma, and has previously been described in terms of energy density. On the other hand, radiation losses ($E_{rad}$) are those which result from the motion of a charged particle under the influence of an electric field. Both are as a result of the energy supplied ($E_{in}$) and hence can be expressed in terms of it as follows (see fig. 13):

$$\mathbf{\eta_{in}\, E_{in} = E_{rad} + E_{th}}$$

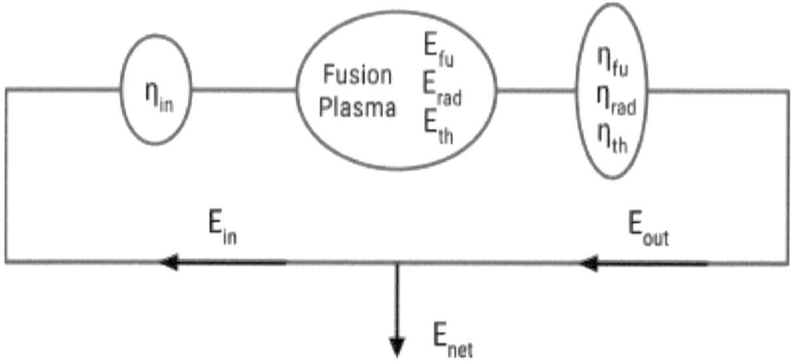


Figure 13: Schematic of the energy transformations of a fusion reactor

It should be noted that $E_{th}$ and $E_{rad}$ amount to a fraction of $E_{in}$ $\eta_{in}$ (coupled to plasma) Thus, only the fraction of $\eta_{in}$ $E_{in}$ is deposited to the net energy to sustain fuel and maintain a constant temperature. In light of this, energy viability can be written now as:

$$(\eta_{fu} E_{fu} + \eta_{rad} E_{rad} + \eta_{th} E_{th}) > (E_{rad} + E_{th}) / \eta_{in}$$

The left side comprises all the recoverable energy, while the right represents the energy supplied. If an idealised system is to be assumed, with global energy confinement time $\tau_E$ that produces power at a constant rate, bremsstrahlung radiation accounting for all the radiation present, the number of ions being equal to the number of electrons, the temperatures of different particles approximating each other such that $T_i \approx T_e = T$, the distribution of particles in plasma volume V being homogenous, and a 50:50 fuel mixture ($N_a + N_b = N/2$), then the well-known Lawson criterion for energy viability can be given, with $<\sigma v>_{ab}$ as a function of temperature, as well as $\eta_{(\ )}$, $Q_{ab}$, and $A_{br}$=Kronecker's constants, as:

$$N\tau_E > [3(1-\eta_{in}\,\eta_{out})\,T] \,/\, [\eta_{in}\,\eta_{out} <v>_{ab}(T)\,Q_{ab} - (1-\eta_{in}\,\eta_{out})\,A_r\,\sqrt{T}]$$

Evidently, The product of the number of particles and the time for which they are confined must exceed a particular function of ion temperature for energy viability. The Lawson criterion suggests that $\eta_{in} \times \eta_{out} \approx 1/3$. This would imply that the lower bound of this product as a function of plasma temperature can be shown graphically as such:

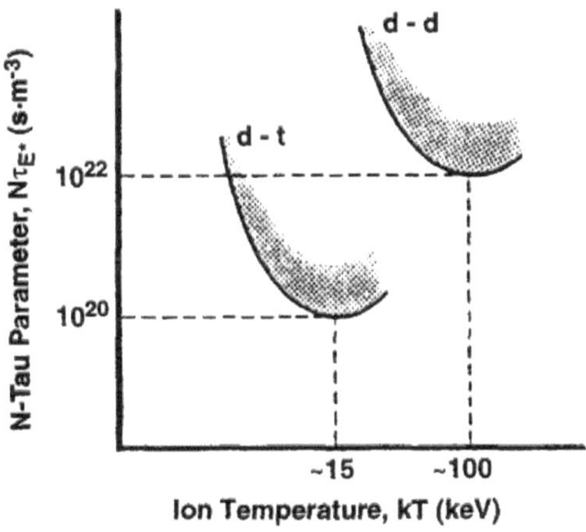


Figure 14: Lawson criterion bounds for d-t and d-d fusion

There exists a point in temperature at which NτE has a minimum value. This can be seen to vary for D-T and D-D fuel, with the former having this value at the point of roughly 15 keV (with $N\tau E = 10^{20} m^{-3} s$), whereas for the latter it is around 100 keV, signifying that D-T reactions can occur under much lower temperatures and hence, are more favourable. The conditions necessary for breakeven are defined by a combination of particle densities, confinement times, and the temperature needed for them to undergo fusion reactors at a sufficient rate. It should be noted, however, that this criterion doesn't pertain to any particular reactor design but serves as a useful parameter for all, as it describes the fundamental characteristics of a fusion plasma that are necessary to produce breakeven and ignition conditions. For instance, even though a direct drive inertial confinement reactor operates on a very fast burn cycle, by considering the time from which the beams make uniform contact with the fuel pellet until the thermonuclear burn process renders the pellet blown apart the burn time, it is possible to account for it in the same way as a magnetic confinement system. Thus, the Lawson criterion remains to be one of the most broadly applicable parameters in the assessment of fusion reactors.

Earlier numerous assumptions were stated regarding the state of the plasma, and hence, even if a reactor is able to exceed the Lawson limit, then it is not yet necessary that it has achieved a state of breakeven. This is due to the fact that it doesn't account for cyclotron radiation emission and the difficulties associated with sustaining the operational parameters during the time interval $\tau_E$ in a manner that they remain constant (as denoted in the equations). This, coupled with the fact that a fusion reactor must produce significantly larger amounts of power than it consumes in order to compete with other energy sources means that for commercial power applications, the Lawson limit must be exceeded by a factor of ten or more.

## Ignition and Break-Even

Previously in the chapter the possibility of a fusion reactor to achieve a state of ignition was considered. However, it is important to reconcile this with the Lawson limit for breakeven in order to arrive at a better understanding of the conditions necessary for a reactor to achieve sufficient temperature to first break even and then proceed to produce energy. In this case, a reactor can be considered to be energetically viable and is characterised by a state in which external power is no longer delivered, ie. The fusion reactions can support themselves. And if the condition for breakeven can be achieved with external heating methods, then the energy used for them now becomes energy gained from the reactor. Such an ignited state is characterised by:

$$F_{c,dt}\, P_{dt}\, \tau_E \geqslant (P_{br} + P^{net}_{cyc})\, \tau_E + 3NT$$

Here it is assumed that $N_i = N_e = N$ and $T_i \approx T_e = T$. Assuming a 50:50 split of deuterium and tritium, ie. $N_d = N_t = N/2$, and substituting the power terms with density-temperature dependant expressions, the equation can be re-written as:

$$(N\tau_E) \geqslant 3T / \{f_{e,dt}\, Q_{dt}\, [(<\sigma v>_{dt}(T)) / 4] - A_{br}\sqrt{T} - [(A_{cyc}\, B^2 \sqrt{T}) / N]\}$$

This represents a new fusion plasma ignition criterion, though doesn't contain energy conversion efficiencies. The relationship between plasma temperature T, measured in keV, and this so-called $N\tau_E$ parameter, with units in $10^{20}$s x $m^{-3}$, can be depicted graphically as so:

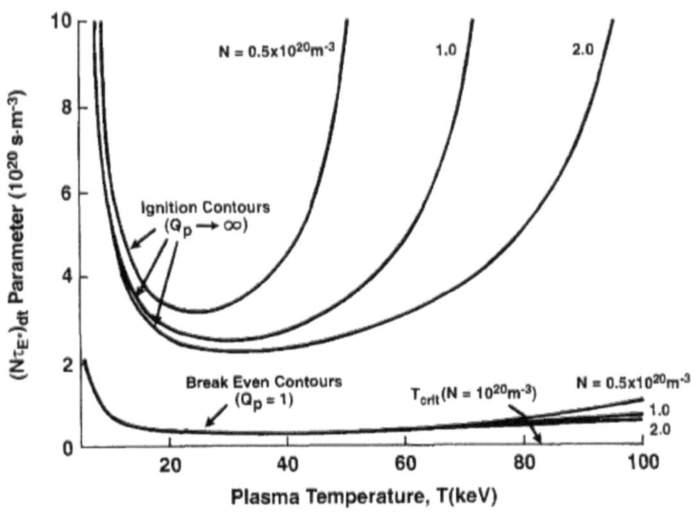


Figure 15: Ignition and break-even conditions for d-t fusion

This graph closely resembles that of the Lawson criterion from the previous section. However, there are a few significant differences. The minimum temperature for ignition for $N=10^{20}m^{-3}$ is seen at T ≈ 30 keV, necessitating $\tau_E$ = 2.7 seconds and $N\tau_E \approx 2.7 \times 10^{20}$ (product of global energy confinement time and plasma density). $N\tau_E$ density dependence is observed due to $P^{net}_{cyc}$ being accounted for with ψ = 0.01.

Interestingly, the denominator in the equation can assume a negative value when the temperature is sufficiently high. Such temperatures can come about due to the domination of cyclotron radiation in the plasma energetics. A negative plasma energy balance implies that the plasma cannot be ignited, and in such a regime the equation does not apply. However, when the plasma temperature approaches a certain critical value $T_{crit}$, then the ignition contours on the graph tend towards infinity, ie. The plasma has reached a point where a feedback loop is created between fusion fuels and products such that there is no longer a need for external heating. Although, this means that the requirements for ignition become increasingly stringent as the temperature increases. An example of such a critical value is when $N = 10^{20}m^{-3}$. It would be useful to consider another way to define fusion reactor performance- in terms of scientific (energy) breakeven given by:

$$\text{Fusion energy output / Plasma Energy Input} = E_{fu}/\eta_{in}E_{in} = Q_p = 1$$

Here $E_{fu}$ amounts to a magnitude equal to the product of $\eta_{in}E_{in}$, making the process neither gain nor lose any energy. The term $Q_p$, then, must be equal to 1 for the net energy to be zero, and approach infinity for the reactor to achieve ignition. If the plasma energy input is zero and yet there is a positive integer value for fusion energy output, then the solution would be infinity ($E_{fu}/0$). While, as it stands, the $N\tau_E$ parameter is very useful, the dependance on both a sufficiently large $N\tau_E$ and temperature for a viable fusion regime signifies the need for the use of the product of the two $TN\tau_E$ in order to accurately qualify the achievements of experiments. This triple product can be obtained by multiplying T on both sides of the equation. This doesn't provide any new characteristic, and the dependence on temperature is similar to what it is in the graph above.

# CH 3: Magnetic Confinement

"The confinement of plasma by means of magnetic fields, as sketched here, may bring the hope of achieving controlled thermonuclear reactions into the realm of practical possibilities."

- Lyman Spitzer

Magnetic Confinement drives are of principally two types, those with open field lines and those with closed ones. What open and closed mean here is evident in the context of fusion reactors- the former implying that the field lines extend out from the confinement region and intersect with the walls of the device, while the latter refers to a state in which field lines are of a continuous, closed-loop, typically achieved by defining the plasma in a ring shape. Some examples of open confinement devices include magnetic mirrors and pinch devices, both of which will be considered, whereas some examples of closed devices include tokamaks and stellarators. Evidently, the latter method of magnetic confinement is considered a more practicable means of fusion plasma confinement. However, the principles behind these systems have been drawn from research in open magnetic confinement, and hence, its study provides useful insights into the fundamental limitations of using magnetic fields to confine plasmas. Moreover, it is essential to recognize in what ways open systems fall short so that future magnetic confinement devices are not prone to face the same challenges. Thus, this chapter will begin with the consideration of open magnetic confinement.

# Open Magnetic Confinement

## Magnetic and Kinetic Pressure

The behaviour of plasma on a macroscopic level is not definable due to the internal field that is present even in straight magnetic field lines. Any form of plasma containment is time-independent due to the associated plasma state having a balance of identified forces, ie. A precondition for successful confinement with magnetic fields, for the most part, is the organisation of the plasma in such a manner that the reactor can operate in a steady state. Thus, the following equations of pressure gradient force and electric current density don't account for the time by assuming that the described conditions are present throughout the burn time. For an isotropic case $(v_i \times \nabla)\, v_j = 0$, then, in the absence of an electric field:

$$\nabla p_j = p_j^c v_j \times B$$

Thus, the total plasma pressure gradient force can be given by:

$$\begin{aligned} \nabla p_j &= \nabla (p_i + p_e) \\ &= (p_i^c V_i + p_e^c V_e) \times B \\ &= j \times B \end{aligned}$$

The equation could be re-written in this way on the basis of the assumption that j= i= ions and j=e=electrons. Here j is the electric current density, and it must satisfy Maxwell's equation:

$$j = \nabla \times (B / \mu_0)$$

The equation for the total plasma pressure gradient force, then, suggests that in a state of equilibrium, the pressure gradient force is equal to the Lorentz force. However, both the electric current density j and magnetic field strength B act in a direction perpendicular to that of the plasma pressure gradient force ∇p (see fig. 16).

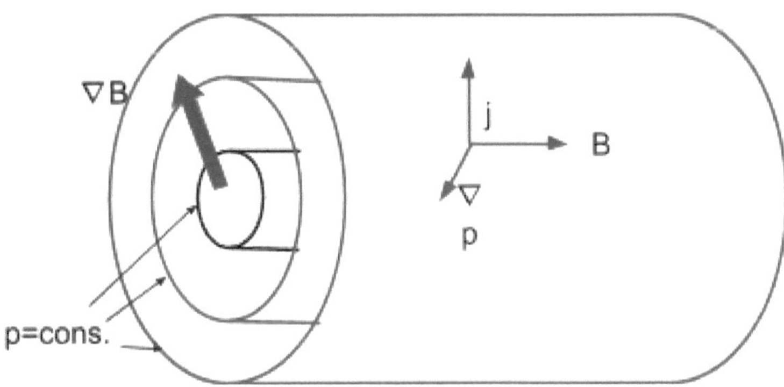


Figure 16: Perpendicular directions of B, j, and ∇p in a closed magnetic system.

Note that j and B lie on an isobaric surface, that is, they are along a surface of constant pressure. In light of the relationship between ∇p, j, and B, this means that ∇p is perpendicular to the surfaces where pressure is constant everywhere. It is possible to combine the equation of the plasma pressure gradient force and Maxwell's equation from above, making ∇p given by:

$$\nabla \mathbf{p} = (\mathbf{1}/\mu)(\mathbf{B} \times \nabla)\mathbf{B} - (\mathbf{1}/\mathbf{2}\,\mu_0)\,\nabla \mathbf{B}^2$$

Assuming that the field lines are straight and parallel to each other, the equation can be simplified and subsequently re-arranged to show (with H denoting local field strength) that the sum total of kinetic pressure and magnetic field energy density remains constant within the plasma:

$$\nabla [p + (B^2 / 2\mu)] = 0$$

$$p + (B^2 / 2\mu_0) = \text{Constant}$$

$$(E^*_{mag} / v) = (BH / 2) = (B^2 / 2\mu_0)$$

What's interesting is the fact that $\nabla p$ induces a current in the direction shown in the figure above. However, this induction causes the magnetic field strength to decrease in the plasma. In addition, the pressure gradient force can result in a particle drift velocity given for a fluid element with n number of particles by

$$v_D, \nabla p = - [(\nabla p \times B) / (N_q B^2)]$$

Accounting for the difference in the drifts of ions and electrons as a result of their properties, it would be necessary to write the equation for electric current density, also called the diamagnetic current, as:

$$j = N_i q_i v_{D,i} + N_e q_e v_{D,e} = (B \times \nabla p) / (B^2)$$

That the direction of the diamagnetic current is perpendicular to that of the strength of the magnetic field is evidenced by cross multiplying the definition of current density as per the pressure gradient equation with it. By supposing the behaviour of particles to be in accord with ideas of gas laws (pv = N* (total number of particles) kT), the total kinetic pressure of an ensemble of ions and electrons can be given by:

$$p = (N^*_i / v) kT_i + (N^*_e / v) kT_e$$

With the assumption that the temperature of the electrons and ions is identical for a Maxwellian distribution ($T_e = T_i = T$):

$$p = N_i\, kT_i + N_e\, kT_e$$
$$= (N_i + N_e)\, kT$$

Recall that these terms were used to define the Beta parameter- the ratio of kinetic pressure to magnetic pressure- which can be given by:

$$\beta = [(N_i + N_e)\, kT \,/\, (B^2 \,/\, 2\mu_0)$$

Evidently, then, what characterises a magnetic system is the interaction or coupling between the kinetic and magnetic pressure. The nature of this interaction is complex, partly due to the fact that the electric field induced by the pressure gradient force reduces the strength of the magnetic field. Moreover, there exists a limit beyond which magnetic pressure can no longer be increased due to the temperatures of the plasma enabling particles to escape, and hence, as a system approaches this limit the degree to which an increase in magnetic field strength corresponds to pressure changes. For the time being, however, a general relation of pressure can be expressed as

Pressure gradient $\propto$ 1/ particle pressure density/mass density

However, if the strength of the magnetic field were to have a minimum value of zero due to the diamagnetic currents generated by pressure gradient force, then the value of β would be infinite since zero would be in the denominator. Evidently, this is impossible, so it is necessary for the definition of the Beta parameter to be altered as the ratio of maximum particle pressure to maximum magnetic pressure. In the case of the diagram above, this is the ratio of the centre of the plasma cylinder to the outer surface of the cylinder. The value of β, then, is limited to being greater than or equal to 1. A high β is desirable, as it signifies that a state of high particle pressure is maintained with a weaker magnetic field/lower magnetic flux, ie. less energy is expended in obtaining the present state of the plasma. Moreover, fusion power density is proportional to $\beta^2$. Nevertheless, there remains a limit to how high β should be, as, beyond a point, instabilities arise in the plasma which disrupts confinement.

## Magnetic Flux Surfaces

The motion of a particle in a simply axially symmetric B-field in a cylindrically symmetric configuration, with j representing the poloidal current, is depicted below:

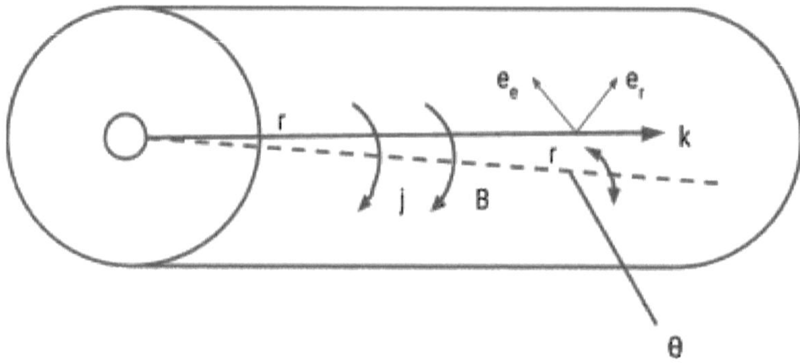


Figure 17: Particle motion in a cylindrical confinement device

It is possible to represent B in terms of vector potential A as suggested by electromagnetic theory by the following:

$$\mathbf{B} = \nabla \times \mathbf{A}$$

Because the currents which generate the field are in the poloidal direction $e_\theta$, A possesses $A_\theta$, its only non-zero component. Thus, B can be represented as

$$\mathbf{B} = \nabla \times \mathbf{A}_\theta\, \mathbf{e}_\theta = -\,[(\partial \mathbf{A}_\theta) / \partial_z] + (1 / r)\, (\partial / \partial r)\, (r\mathbf{A}_\theta)\, \mathbf{k}$$

In the absence of any other external field, the equation of particle motion can be given by:

$$m(dv / dt) = q\, \big(\mathbf{E}_A + \{\mathbf{v}_x[-(\partial \mathbf{A}_\theta / \partial_z)_{er} + (1 / r)\, (\partial / dr)\, (r\mathbf{A}_\theta)\mathbf{k}]\}\big)$$

Here $E_A$ represents the electric field consistent with the temporal variation of B (satisfying Maxwell's relation), allowing for the following equation and the conditions which make it satisfied to be framed respectively as

$$\mathbf{B} \times \nabla(r\mathbf{A}_\theta) = \mathbf{B}_r\, [(\partial(r\mathbf{A}_\theta) / \partial r] + \mathbf{B}_z\, [\partial(r\mathbf{A}_\theta)/(\partial_z)] = 0$$

and

$$\mathbf{B}_r = \mathbf{e}_r\, (\nabla \times \mathbf{A}) = -[(\partial \mathbf{A}_\theta)/(\partial_z)]$$

$$\mathbf{B}_z = \mathbf{k}\, (\nabla \times \mathbf{A}) = (1/r)\, (\partial/\partial r)\, (r\mathbf{A}_\theta)$$

In the absence of any other forces, the guiding centres of the particles move on surfaces where $rA_\theta$ = constant. These are called the flux surfaces of the magnetic field. Interestingly, symmetry here constrains all gradients into the radial direction and drifts into poloidal directions. A consequence of this is the fact that in a cylindrically symmetrical plasma, in spite of the particle drifts which occur across the lines of the magnetic field due to the existence of variations in its strength, particles would remain on their respective flux surfaces. This phenomenon evidently makes flux surfaces play an important role in the open and closed systems later considered.

## Magnetic Mirror

A magnetic mirror is the simplest method for obtaining a cylindrically homogeneous magnetic field. In such a device, a current-carrying solenoid is used to trap charged particles in a cylindrical chamber. Under normal circumstances, fuel ions, and hence thermal energy, would be lost due to the plasma leak at the open ends. However, by increasing the strength of the magnetic field at the ends, the ions are not only kept from making contact with the reactor walls but are reflected from the regions in which the field strength is at its maximum (see fig. 18). The mechanism by which this occurs is the constraint on the motion of an ion, mainly its conservation of energy and magnetic moment.

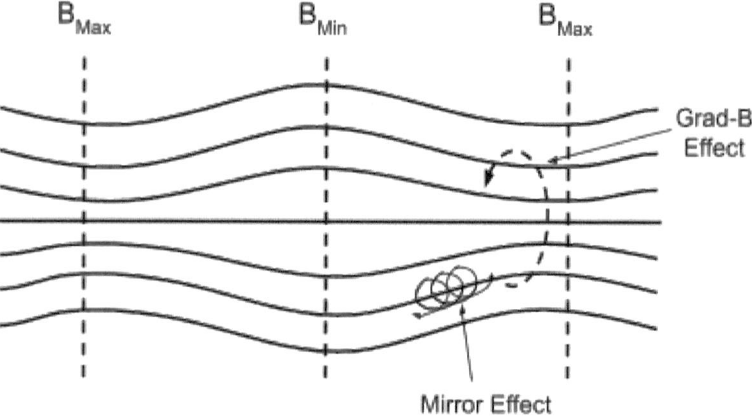


Figure 18: Magnetic mirror with helical particle reflection along field line

It would be possible to describe the features of such a device in terms of the behaviours the ions trapped within it demonstrate as a result of moving in both a homogeneous and inhomogeneous magnetic field. Firstly, as previously established, the basic motion of an ion in such a scenario entails a counter-clockwise spin such that the centre of gyration moves in the direction of a magnetic field line, ie. it spins around as well as in the direction of the field line (see fig. 19).

Figure 19: Particle Motion Around Field Line

Drifts occur in azimuthal directions, however, and particles remain bound to magnetic surfaces in such a manner. Moreover, ions experience drifts when approaching the ends of the confinement chamber due to the inhomogeneity of the field. These drifts turn into losses in the direction parallel to the magnetic field. Predominantly, though, these ions are confined in relation to the radial direction (perpendicular to B), with the only exceptions being attributable to the migration of plasma outside the contained region due to collisional and turbulent transport causing diffusion in the perpendicular direction. A possible method for reducing such losses is to squeeze the b-lines, that is, increase the strength of the magnetic field. The degree to which such a method offers a viable solution will be discussed

First it is important to discuss how the constancy of the magnetic moment and the conservation of the energy of particles prevents them from escaping the magnetic "bottle" of the mirror field. Firstly, the parallel and perpendicular velocity components of a particle established by the magnetic field lines in terms of its energy, with $E_0$ denoting a constant of motion in the absence of the effects of other forces, is given by

$$\frac{1}{2} m v_{//}^2 + \frac{1}{2} m v_{\perp}^2 = E_0$$

Since $E_0$ = constant, it is possible to depict the kinetic states of ions within a collisionless plasma on a line in the $v_{\perp}^2$ - $v_{//}^2$ plane as follows:

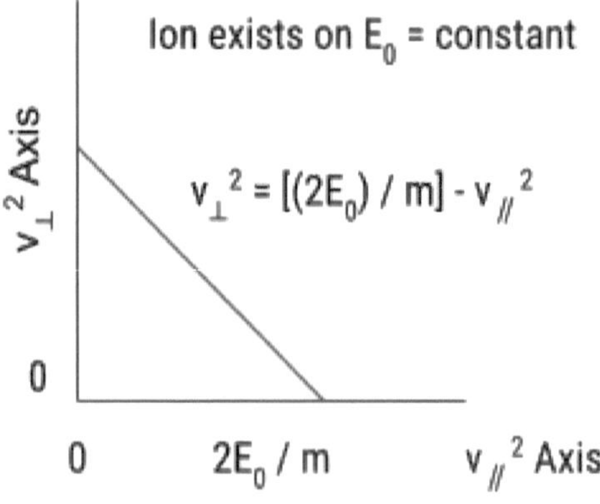


To account for the conservation of energy among the particles the term μ (magnetic moment):

$$E_0 = \frac{1}{2} m v_{//}^2 + \mu B$$

Based on the position of the ion in the mirror, the strength of the magnetic field it experiences will vary between $B_{min}$ and $B_{max}$ whether it is towards the centre or the ends of the solenoid respectively. Evidently then, as the sum of the parallel and perpendicular velocity components comprise the total energy of a particle, the fact that $v_{//}$ is at a maximum when B is at a minimum (the centre of the mirror), implies that the opposite is true as well- when B is at a maximum $v_{//}$ is at a minimum

$$v_{//} \text{ Maximum} \rightarrow \text{B Minimum}$$
$$v_{//} \text{ Minimum} \rightarrow \text{B Maximum}$$

In light of this, the range of magnetic field variation can be introduced:

$$E_0 = \frac{1}{2} m (v_{//}^2)_{max} + \mu B_{min}$$

From this, as well as the corollary that the particle motion in the central plane is towards the parallel direction whereas it is towards the perpendicular direction near the ends, it is possible to define a condition for trapping particles (prohibiting escape through the ends by stopping their parallel motion at the $B_{max}$ plane), in the absence of collisional effects, as:

$$v_{//} |_{B \leq B\,max} = 0$$

This can be combined with the previous equation, making the condition in terms of constant of motion in the absence of the effects of other forces $E_0$ as:

$$E_0 \leq 0 + \mu B_{max}$$

Hence, trapping requires, with the equality sign yielding the maximum value $v_{//}^2$ can assume, the following:

$$\frac{1}{2} m(v_{//}^2)_{max} + \mu B_{min} \leq \mu B_{max}$$
$$[\frac{1}{2} m (v_{//}^2)_{max}] / (\mu B_{min}) \leq (B_{max} / B_{min}) - 1$$

The definition of magnetic moment can then be applied in order to arrive at a relation for the speed components at the midplane:

$$(v_{//}^{2})_{max} / (v_{\perp}^{2})_{min} = (v_{//}^{2} / v_{\perp}^{2})_{mid\text{-}plane} \leq (B_{max} / B_{min}) -1$$

This evidently suggests the intuitive fact that the parallel component of the speed of the ion should be relatively small so that its chance of escape is diminished. This is since, if an ion were to have a sufficiently high velocity in a direction parallel to the field lines, then it would pass through the ends. However, if it were to travel in a perpendicular direction, then it would not exit the container, but rather simply approach the inner walls. This explains the principle on which the condition above is established- that an ion's parallel velocity component increases as the strength of the magnetic field decreases. The field line the centre of gyration of the ion follows leads to the ends of the mirror, and hence, only at instances where the field is weak is the ion following such a trajectory. Approaching the ends, however, the field prompts the ion to adopt a perpendicular velocity component, thus, due to the conservation of energy, necessitating a decrease in the parallel component.

The probability of a particle being trapped or lost through the ends is determined by the relative length of the two sections of the $E_0$ = constant line on the graph above in the following manner:

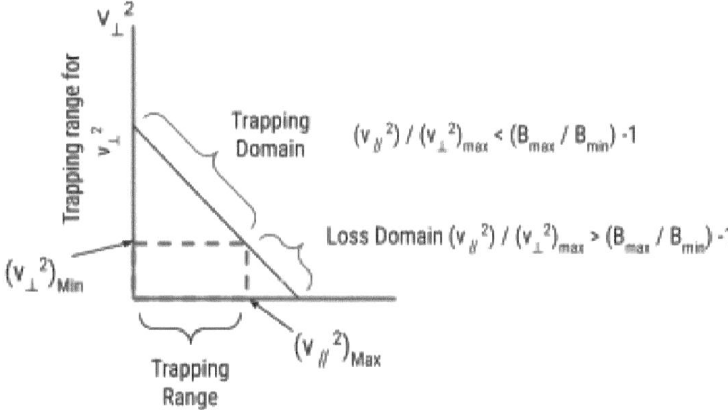


The ratio of the parallel and perpendicular velocity components identifies a domain on E = constant line with which it is possible to associate the ranges of possible values $v_{//}^2$ and $v_\perp^2$ can assume that will result in the ion being trapped. In addition, it is possible to determine the direction in which a particle must travel so that it remains trapped by considering its motion in velocity space as shown below:

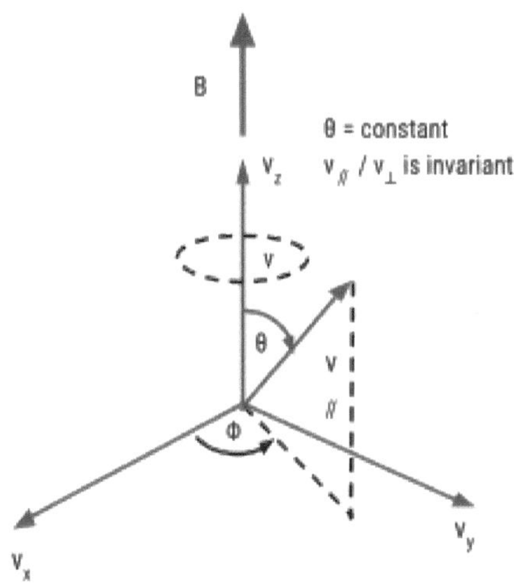


Figure 20: Particle Motion in Velocity Space

Based on such a particle motion, it is possible to frame the following relation and subsequent re-arrangement:

$$v_{//} / v_{\perp} = \cos\theta / \sin\theta$$
$$\sin^2\theta = 1 / (v_{//}{}^2 / v_{\perp}) + 1$$

The latter reveals the fact that particles travelling in a region of the mirror that experiences a relatively weaker magnetic field (roughly the mid-plane), are trapped. Consequently, those which are travelling in the so-called "loss cone" escape from the plasma region. The condition for each under which they

What constitutes these motions can be expressed, for the particles trapped as a trigonometric function and those lost as a polar angle, respectively by

$$\sin\theta \geq \sqrt{B_{min}} / B_{max}$$
$$\theta_0 = \sqrt{B_{min}} / B_{max}$$

This signifies, as mentioned above, that as particles move infinitesimally closer to a region in which the strength of the magnetic field is greater, they would seem to violate the law of the conservation of energy. This is due to the fact that a displacement of this sort would result in the particles gyrating at a faster rate and thus, increasing their perpendicular velocity component to the point where $v_{\perp} > (v_{\perp})_{max}$. Evidently, then, additional energy is transferred in the form of the particle moving from the plane at which it possesses the maximum perpendicular velocity component to the plane associated with $v_{\perp} < (v_{\perp})_{max}$. And, since, as previously established, the sum of the parallel and normal velocity components comprises the total kinetic energy of a particle, the movement in this direction allows for a gain in $E_{//}$. Moreover, because $v_{\perp} < (v_{\perp})_{max}$ requires a weaker magnetic field, the only way in which a particle can regain $E_{//}$ is by moving the opposite to the gradient of the magnetic field. Thus, the particle is reflected and returned to a region with a weaker field.

The determinations of the relative requirements in terms of magnetic field strength and plasma pressure for effective containment made previously assumed a state of magnetohydrodynamic equilibrium. That is, it was assumed that the solutions to the magnetohydrodynamic equations were time-independent such that the state of the plasma could be sustained indefinitely. In reality, however, that this state would not remain stable was not considered. In a stable equilibrium, where any small perturbations that occur are inherently dampened, the associated forces are balanced and hence, a steady-state solution of the MHD equations can be achieved. The requirement for such an unperturbed state is perfect thermodynamic equilibrium within the plasma. This means that the plasma particles must have a Maxwellian distribution, as well as maintain a uniform density. Moreover, it is necessary that the magnetic field is uniform as well. However, in the context of the magnetic confinement configurations that are relevant to nuclear fusion, the above requirements are not satisfied.

Consequently, the equilibrium is unstable. That is, small deviations from the initial equilibrium state are amplified by the propagation of perturbations with time (instabilities). For magnetic mirrors, this occurs due to the loss of particles through the so-called “throats” of the mirrors. Further, this loss disrupts the isotropy (uniformity in all directions) of a Maxwellian distribution, since the escaped particles possess a dominant $v_{//}$ component. Moreover, such a state features both $\nabla B$ and $\nabla N_i$, as the strength of the magnetic field varies, while the ions, due to how their motion under the influence of a magnetic field differs from that of electrons, become concentrated in certain regions. A consequence of this is the fact that the reaction rate significantly decreases due to the large accumulation of like charge- the coulomb barrier between ions in the absence of electrons is too high for them to get in sufficient proximity of one another.

It is possible to balance all the forces acting in a steady-state reactor. However, the state would then possess so-called "free energy", which can give rise to instabilities due to the lack of a perfect thermodynamic equilibrium. A consequence of this is the possibility of the occurrence of periodic motion of the fluid elements of a plasma akin to oscillations or waves. An instability, then, refers to the motion of particles within a plasma that brings it closer to a perfect thermodynamic equilibrium by reducing the quantity of free energy. The exact nature of such instabilities varies based on the reactor configuration. Nevertheless, the instabilities of magnetic mirrors share characteristics with those that plague other magnetic confinement devices, and hence, their consideration reveals something broadly about the disposition of instabilities. Specifically, flute-type instabilities are the ones most relevant to mirror machines, and hence, will here be discussed.

Flute-type instabilities describe a state in the plasma wherein 'flutes' of plasma move across magnetic field lines as such

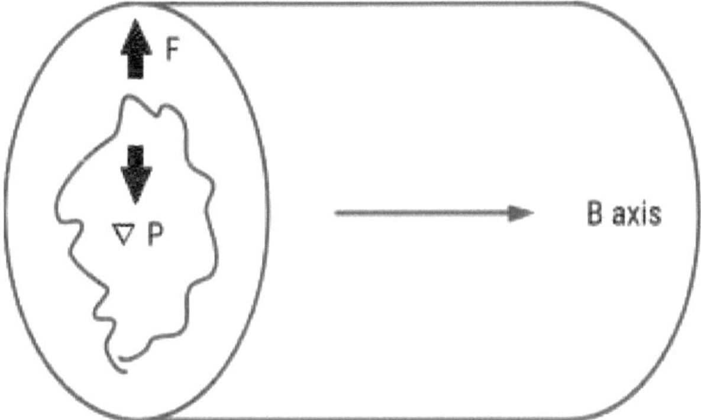


Figure 21: Flute of plasma in a mirror device

This occurs due to the displacement of the fusion plasma into a state of lower magnetic induction and kinetic pressure. The movement of the plasma to regions with lower energy density allows for the kinetic energy of particles to allow the perturbation to grow, ie. in locations wherein the particles aren't travelling with a very high kinetic energy, those which move in response to the flute-type instability are more easily able to impart their motion to their surrounding in the form of collisions, thus, allowing the instability to grow. The reason for why plasma in mirror devices does this appears to be due to the energy density gradient present on its boundary surface when confined magnetically. Consequently, the displacement mentioned above that is responsible for transporting the plasma to lower energy density regions occurs as a result of outward perturbations or ripples on this surface. This means that such instabilities are unavoidable with such a device, as it is impossible for the plasma approaching the boundary surface to not experience an energy density gradient due to its position in the device. In the mirror geometry of the figure above, this phenomenon can be observed in the manner in which the curvature of the magnetic field is seen to be convex everywhere with the exception of the end regions. The movement of the plasma in such a manner causes particles to be lost from the region of the reactor beyond which they are no longer contained.

Evidently then, what constituents an unfavourable convex curvature of the magnetic field is where curvature drift- when particles move parallel to the field lines rather than gyrating around them- leads to charge separation in the plasma in a direction perpendicular to the magnetic field and the radius of curvature, ie. electrons and ions forming clusters of like particles in distinct sections of the mirror as shown below

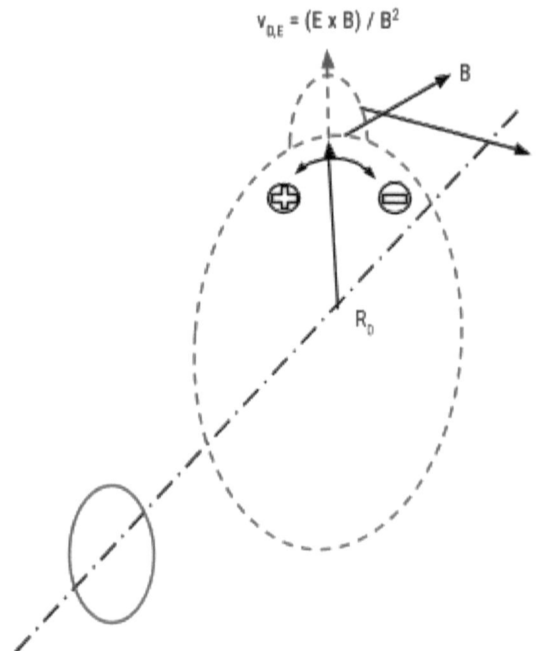


Figure 22: Curvature drift resulting in charge separation

Here curvature drift leads to a state of azimuthal polarisation (polarisation that is perpendicular to a cylinder's radial axis), creating the electric field the interaction of which with the magnetic field gives rise to a phenomenon known as E x B drift. This displaces both ions and electrons in the radially outward direction. This transportation of particles is what is responsible for the flute-like bumps on the plasma column- groups of particles are moving towards the boundary of the plasma as a result of the drifts. However, this drift should not be confused with that which causes azimuthal polarisation. In this regard, it would be useful to summarise the process by which a plasma confined n a mirror device can give rise to a flute-type instability:

1) A convex curvature of the magnetic field results in particles moving parallel to the magnetic field lines rather than gyrating about them (curvature drift)
2) This compounds in a state of azimuthal polarisation, wherein charge separation occurs at normal angles to both the magnetic field as well as the radius of curvature
3) The polarised state creates an electric field within the plasma
4) The electric field, by interacting with the magnetic field, gives rise to another set of drifts known as E x B drifts
5) Particles are displaced radially outward by the E x B drift

The shape assumed by the plasma in the diagram then, of flute-like bumps on the plasma column, can be explained by the fact that particles are moving parallel to the magnetic field, and thus, creating the conditions under which an electric field is generated, however the resulting E x B drift, by virtue of the interaction between this generated field and the shape of the already present magnetic field, is causing the displacement of particle in some regions more than others. The way to avoid this instability is to create a 'minimum-B' field configuration. That is, one in which the field lines are almost concave everywhere into the plasma. Here minimum-B refers to a magnetic topology within which charged particles find themself in a magnetic well, as it appears that the strength of the magnetic field is increasing in every direction, thus, prompting the particle to stay where it is. A similar concept is applied in inertial electrostatic confinement reactors, only using an electrostatic well rather than a magnetic one. The way this works is by curving the structure of the confinement chamber in such a manner as to give a particle the impression, so to speak, of being surrounded by a stronger magnetic field in the direction in which it can move. An example of such a device is one shaped like a baseball seam, which, when suitably flattened and oriented in opposition with another similar coil, becomes what is called a "yin-yang" coil configuration.

There exist many variations of the minimum-B concept, but all are based on providing a central, circular region within which the plasma is contained. Moreover, such devices are characterised by having, flattened, fan-shaped magnetic fields in two opposite directions, which, by virtue of being open, still form a magnetic mirror. In principle, it is possible for the magnetic well to be deeper, thus, confining the plasma to a greater extent. More significantly, however, a deeper well allows for such devices to be added to the ends of mirror solenoidal fields. This can also make the particles contained therein represent a thermal barrier that particles trapped in the solenoidal field must overcome if they are to escape.

On the whole, mirror concepts remain unavailable, however, due to the fact that they require a state-state operation, where the rate at which particles are injected into the plasma is identical to that at which diffusion occurs through the open ends. The non-Maxwellian distribution of the particles in a mirror device assumed as a result of the inhomogeneous magnetic field, which causes a disproportionate loss of particles with a large ratio of a parallel velocity component to a perpendicular velocity component, is the principal cause of such diffusion in velocity-space. This is referred to as the 'lost-cone' distribution, which excludes all the particles in the lost-cone. Evidently, this is a significant deviation from the Maxwellian distribution, and hence, gives rise to velocity-space instabilities such as the 'lost-cone' instability- wherein diffusion into the lost-cone is enhanced. However, it is possible to reduce the effect of such instabilities on the confined plasma by reducing the dimensions of the mirror devices in which they are contained. Here is a feature distinct from other magnetic confinement concepts, which require a larger diameter in order to prevent curvature drifts from disrupting confinement. This is a result of, among other factors, the unique interaction between the electric and magnetic field that is responsible for ExB drift, as well as the fact that the progression from an increase in surface area to that of volume is not linear but rather geometric. Needless to say that scaling laws add a limit to this reduction in size, as, beyond a point, it is impossible to create conditions of sufficient temperature and pressure due to the fixed size of atoms.

### Classical Mirror Confinement

If the role of collisions is to be neglected, then all particles which don't appear in the loss cone in velocity space are trapped. Yet, collisions randomly bring ions and electrons into the loss cone from the region in which they would otherwise be confined. Within a single transit time $\tau = L/v$ over the length L of the device, particles escape the device once they have entered the loss cone. Electrons, however, have a greater proclivity to be scattered into the loss-cone and be lost as a result of their smaller mass, and hence, diffuse more rapidly (in terms of velocity and coordinate space) than ions. This results in the buildup of positive electrostatic potential within the confined plasma. As a result, beyond a certain point of diffusion, the electrons are retained in the magnetic bottle. Thus, overall, the plasma confinement time is based on the time it takes for ions to escape confinement. The ion escape time of a plasma can be characterised by the time of collisions between ions as given by:

$$\tau_{ii} = 1 / N_i <\sigma_s v_r>$$

This equation can be expanded to account for ions encountering multiple collisions simultaneously at plasma temperature which are relevant for fusion, with $A_i$ denoting the atomic mass number of the ions, as:

$$\tau \propto [A_i^{1/2} (kT_i)^{3/2}] / N_i q_i^4$$

Evidently the confinement time in the magnetic mirror $\tau_M$ must be determined by the size of the lost cone, hence, making it approximate the following relation for larger ratios of $B_{max} / B_{min}$ (so-called mirror ratio):

$$\tau_M \approx \tau_{ii} \ln (B_{Max} / B_{Min})$$

Substitution, with $q_i = z_i e$ has been replaced, and c taken to be constant for typical fusion temperatures as $1.78 \times 10^{16}$ s:

$$\tau_M \approx c \{[(A_i^{1/2} T_i^{3/2} \ln (B_{Max} / B_{Min})] / N_i z_i^4\}$$

$T_i$ measured in keV

$N_i$ measured in particle per $m^3$

$\tau_M$ measured in seconds

Mirror confinement time $\tau_M$ depends not on the magnitude of the strength of the magnetic field or the size of the plasma, but rather the average temperature of the ions, and hence, the ratio of $B_{Max}$ / $B_{Min}$. If the fuel ions are more densely packed, then the rate at which scattering into the lost cone occurs increases, and consequently, $\tau_M$ decreases. Note that here the derivation of $\tau_M$ is predicated on the suppression of collective effects such as instabilities, and hence, the confinement time within the mirror, in reality, can be anywhere from slightly shorter to characteristically different based on the injection rate and magnetic topology, among other elements of reactor design.

## Magnetic Pinch

Another popular method of open magnetic confinement involves the use of a magnetic field generated by the plasma itself. This is done by the electric current of the plasma inducing the magnetic field which confines it. Increasing the current results in a larger magnetic field, thus, compressing the plasma. Moreover, this compression raises the temperature by joule-heating. Hence, the energy supplied in the form of the current not only goes towards confinement but heating as well. Evidently, such a process would warrant massive currents- some $10^5$ A. Thus, a pinched device is only operated in short pulsation. There are principally two pinch concepts: the z-pinch and the θ-pinch, both of which will be discussed (see fig. 23).

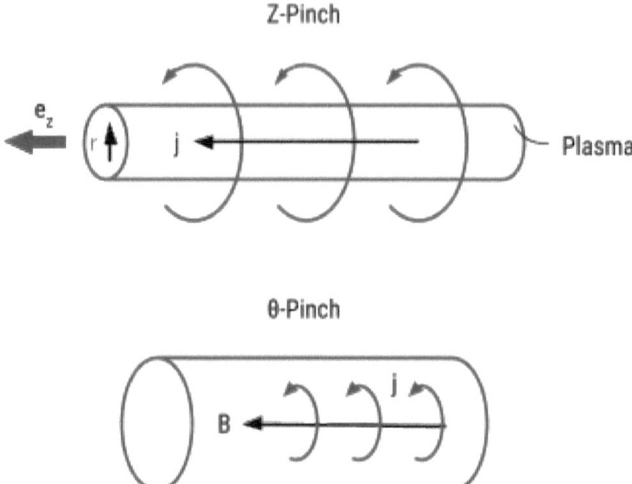


Figure 23: Diagram of z and θ pinch devices

For particles to remain confined in a z-pinch device, it is necessary that the magnetic energy density and plasma pressure are balanced. This can be considered a condition for such devices to operate, and hence, will be considered in context to the required electric current and state of the plasma. Firstly to achieve this balance, a sufficiently strong magnetic field must be generated, hence necessitating a high enough electric current. Such a state can be defined, and subsequently, its magnetic field B induced by the electric field j (determined by Maxwell's equation) be given respectively, assuming $T = T_i \approx T_e$, by

$$\mathbf{B^2 / 2\mu_0 = (N_i + N_e)\ kT}$$

and

$$\nabla \times \mathbf{(B / \mu_0) = j}$$

With this the absolute value of magnetic induction at the surface $B_\theta$ can be given by

$$B(a) = (\mu_0 I) / 2\pi_a$$

Note that B can only possess an azimuthal direction. The insertion of the equation of the value of magnetic induction at surface $B_\theta$ into that outlining the need for the magnetic energy density to match the plasma pressure, yields, if the number of particles is to refer to a plasma column of unit length, ie. in the volume $a^2 \pi \times 1m$ such that $N^* = Ni^* + Ne^* = a^2 \pi (Ni + Ne)$, then what is known as the Bennet Pinch condition can be formulated as

$$\mu_0 I^2 = N^*kT$$

It should be noted that aside from the challenges associated with satisfying this condition, pinch devices are additionally troubled by instabilities. These come in various forms, although considering the case of a cylindrical plasma carrying a large current along its axis, those of interest are the sausage and helical kink instability.

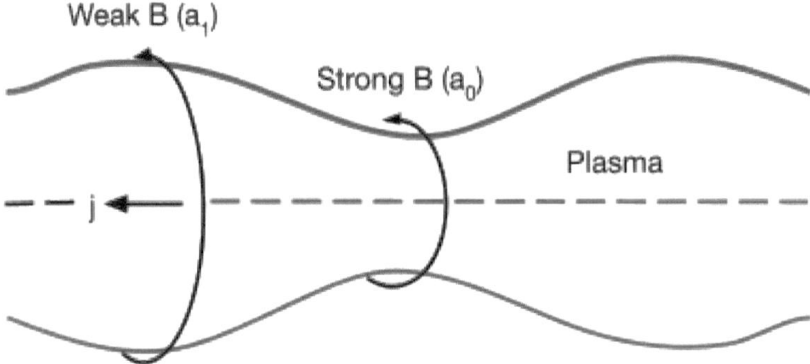


Figure 24: Diagram of a sausage instability

Above is a diagram of a typical sausage instability. It evidently shows that the disturbance of an initial state of equilibrium is by the expansion of the plasma column to a larger radius than that initially considered. That is, it occurs due to axial perturbation in the diameter of the plasma column. Here the magnetic pressure at $a_1 > a_0$ is weaker as compared to the surface of equilibrium $a_0$, and hence, it further reinforces the expansion process due to the fact that $B(a) \sim 1/a$. The occurrence of these perturbations are responsible for making the plasma column resemble a string of linked sausages. Such a state is now preserved, however, as the sausages are disrupted as they grow and more closely approach the reactor walls. A possible solution is to apply a magnetic field of sufficient strength along the axis in order to restabilize the instability.

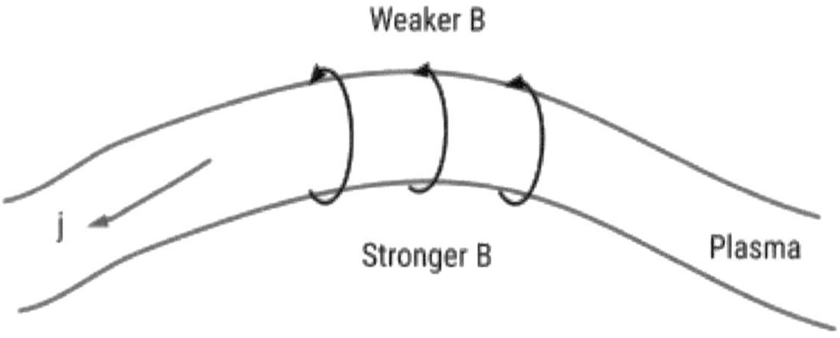


Figure 25: Diagram of a helical kink instability

The diagram provided above depicts the second characteristic instability in a cylindrical pinch device associated with the perturbations of its linear axial geometry. As seen in the figure, this occurs when the plasma column experiences some form of axial curvature, and results in the bend having a larger magnetic induction as a result of the closer proximity of the field lines. The difference between the magnetic pressure inside as compared to outside the bend is responsible for the growth of small disturbances of the state of equilibrium in the plasma cylinder. The bending continues until the plasma column scrapes the surrounding wall, beyond which point containment is disrupted. This unstable response to bending is called the helical kink instability, and can similarly be suppressed by additional strong magnetic fields.

All in all, although a differentiating feature of pinch devices is their ability to generate the magnetic field which confines them, often the instabilities they experience are due to the extreme electrical currents needed for them to do so. Remarkably, however, θ-pinch devices have been shown to be significantly more stable as compared to others. Finally, it is possible to diminish the end losses of a pinch device by adding magnetic mirror configurations. Nevertheless, open magnetic systems, due to the large losses associated with the ends through which plasma can escape, are not favoured as magnetic confinement devices. Instead, current research primarily focuses on closed magnetic systems, those which address the problem of end losses by bending the plasma column into a torus. It is these devices which will now be considered.

## Closed Magnetic Systems

In a closed magnetic field, the field lines don't leave or enter the confinement region, hence, having no ends from which plasma can escape. The simplest method of achieving such a magnetic topology is with the shape of a torus. Such devices comprise the majority of experimental reactors currently in operation and planned for construction, although the cross-sectional area assumes a D rather than a circle. This section will discuss the features and characteristics of such devices in the context of a conceivable energy-generating device.

### Toroidal Fields

The most important feature that characterises a closed magnetic system is the presence of a toroidal field. This field appears doughnut-shaped and is responsible for confining the plasma along a large ring around the torus, encircling the central void. In addition, a poloidal field confines the plasma in a direction that follows a small circular ring around the surface. The combination of these two fields results in the plasma assuming a concentric torus structure in such a manner as to keep the plasma from interacting with the walls of the reactor (see fig. 26). However, before describing their interaction, it is first important to understand the role of the toroidal field by considering it as the only magnetic field present, as this will not only allow for a comprehensive description of its properties but illustrate the need for the poloidal field as well.

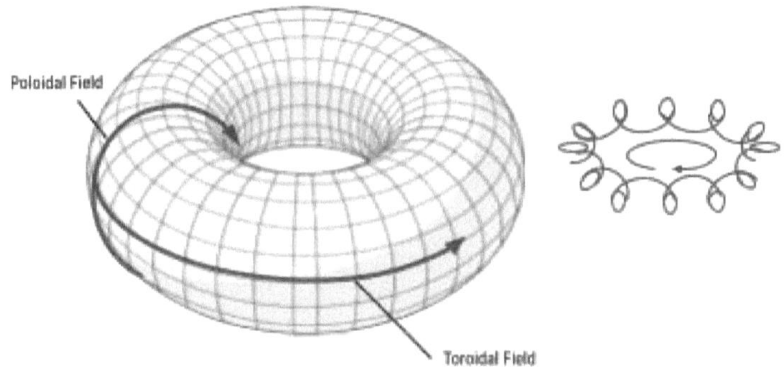


Figure 26: Poloidal and toroidal field in a torus

The toroidal field B is produced by the passage of current through sets of coils wound around the torus. By bending a solenoidal coil around until its ends meet, a straight axial magnetic field can be deformed into an axisymmetric (symmetrical about an axis) toroidal field. Seemingly counterintuitively, the toroidal coils are in the shape of the cross-section of the torus and are placed perpendicular to it, while the poloidal coils circle its major radius. However, this makes sense considering the role of these coils in confining the plasma, of toroidal confinement to separate the plasma from the walls, and for poloidal confinement to separate it from the surface of the poles. This system can be topologically defined by $R_0$ and a, the former denoting the major radius and the latter the minor radius (see fig. 27).

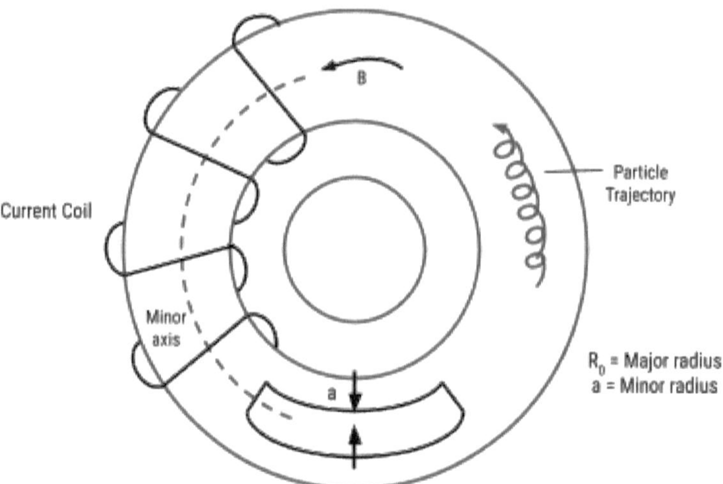


Figure 27: Plasma confined in a torus defined by R0 and a

Note that the major radius refers to the distance between the centre of the torus and the central line between the inner and outer surface, while the minor radius refers to the distance between one of these surfaces and the central line. The variation between the spacing of the two sets of coils greatly affects the magnetic field. When the spacing is closer on the inboard side, the magnetic field experiences a radial variation. The form of radial variation can be given, with R denoting the distance from the major axis, as:

$$\mathbf{B(R) \propto 1 / R}$$

The gradient in the strength of the magnetic field creates what's known as the grad-B force, which is responsible for driving positively charged ions in one transverse direction and negatively charged electrons in the other. Note that curvature drift within the plasma can effectuate this as well. Both of these result in local charge separation, generating a vertical electric field in the process. Moreover, this field causes both ions and electrons to simultaneously drift in outward directions which are perpendicular to the major torus axis. To illustrate, below is a depiction of a section of a torus containing plasma under the influence of a toroidal field. A detailed view of the cross-sectional surface area portrays both the role of the grad-B force and curvature drift in local charge separation, as well as its effect on the motion of particles in upwards directions normal to the torus.

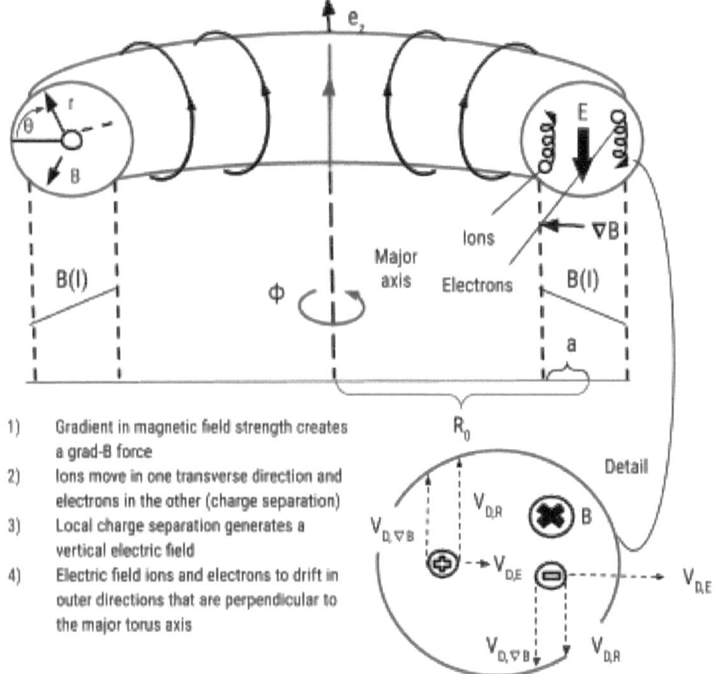


Figure 28: Particle drift in the presence of a toroidal field

To demonstrate, consider the magnetic field only generated from the current in the toroidal coils. In such a case, the field only possesses a component in the toroidal direction $e_\Phi$. Thus, the strength of the magnetic field in such a device can be denoted as:

$$\mathbf{B} = \mathbf{B}_t$$
$$= (\mathbf{B}_{r,}\mathbf{B}_\Phi\mathbf{B}_\theta)$$
$$= \mathbf{B}_\Phi\ \mathbf{e}_\Phi$$

In light of this relation, the magnitude of the toroidal field can then be given, with I as before denoting the current, only now specifically referring to that which is threading the hole of the torus, as:

$$\mathbf{B}_\Phi\ (\mathbf{R}) = \mu_0\mathbf{T}\ /\ 2\pi\mathbf{R} = (\mathbf{R}_0\ /\ \mathbf{R})\ \mathbf{B}_\Phi\ (\mathbf{R}_0)$$

Based on this, it is possible to find an equation to determine the drift velocity the guiding centre of a charged particle moving along a toroidal field will experience, as well as its combination with the previous equations, when said field is denoted by $B_t$, and the parallel velocity component $v_{||}$ is subject to curvature and $0_{\nabla B}$ drift if it posses a components perpendicular to B, respectively as such:

$$\mathbf{v}_0 = \mathbf{m}\ /\ \mathbf{qB}^2\ \{\mathbf{v}_{||}^2\ (\mathbf{R}\ \mathbf{x}\ \mathbf{B}\ /\ \mathbf{R}^2) + (\mathbf{v}_\perp^{\ 2}\ /\ 2)\ [(\mathbf{R}\ /\ \mathbf{B})\ \mathbf{x}\ \nabla\mathbf{B}]\}$$

and

$$\mathbf{v}_0 = (\mathbf{m}\ /\ \mathbf{q})\ [1/\mathbf{R}_0\mathbf{B}_\Phi\ (\mathbf{R}_0)]\ [\mathbf{v}_{||}^2 + (\mathbf{v}_\perp^{\ 2}\ /\ 2)\ \mathbf{e}_z$$

A result of this equation is the polarisation of plasma, with the electric field vector pointing in the negative z-direction. This field, which originates due to E x B drift, causes ions and electrons to move radially outward, with the velocity with which they do so being given by:

$$v_{D,E} = (E \times B^2) = [E / Bq (R_0)] \times (R / R_0)$$

This drifting plasma eventually makes contact with the surrounding wall, and hence, a toroidal field offers little to no plasma confinement due to having practically no radial equilibrium.

## Particle Trapping

A common method to reduce the outward drift of particles which results from charge polarisation is to introduce another field along with the toroidal ($B_\phi$) one. This is the poloidal field $B_\theta$, mentioned earlier as one of the two fields which principally confine the plasma in closed magnetic systems. The combination of these fields on the plasma is the particles assuming a helical migration path based on the relative magnitudes of the individual poloidal and toroidal components. In such a scheme, any undesired charge separation is negated by the particles spending equal time in the upper and lower halves of the toroid as a result of the helical field. Thus, the problem is solved when the particles have moved along a twisted field line through a completed poloidal rotation, $\theta = 2\pi$.

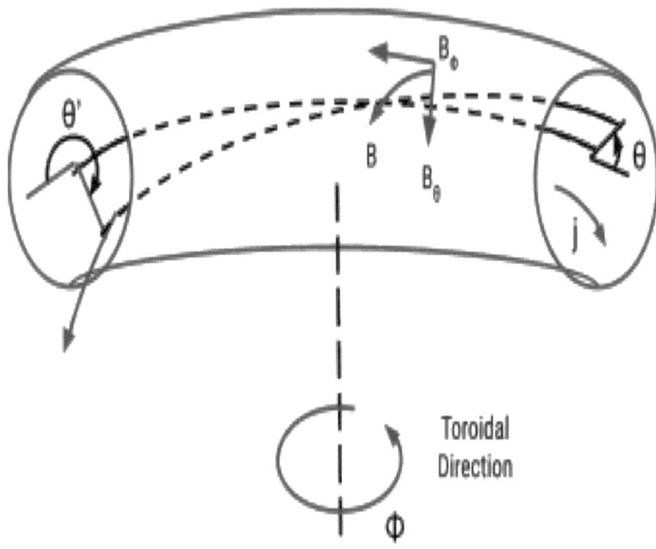


Figure 29: Toroidal and poloidal fields along with the rotational transform

It is now of use to introduce the concept of a rotational transform, represented by $\tau$ (iota), to describe the poloidal angle made by a magnetic field line after its traversal through $2\pi$ revolutions in the $\phi$ direction. When the value of the rotational transform is very large, the revolving field lines confine the plasma tightly enough such that it is unstable against kink-type perturbations. Therefore, a safety factor, indicating how large the rotational transform can be for the reactor to remain viable, as well as the subsequent constrain for the rotational transform, can be defined respectively by

$$q = 2\pi / \tau$$

and

$$\tau < 2\pi$$

This also acts as a measure of the pitch of the field line. Since the poloidal angle to which τ refers is after 2π revolutions, evidently for a plasma to remain stable the value of q must be equal to or greater than 1. However, this requirement is raised to $q \geq 2.5$ for the edge of the plasma. If these requirements are not met, then the plasma lacks the stability to suppress kink-type instabilities once they arise in the plasma. Moreover, put another way, the constraint for the rotational transform is that the number of revolutions around the major torus must exceed the number of revolutions around the minor axis following magnetic field lines. To summarise, the role of the rotational transform is to spatially mix charged particles so that the strength of the vertical electric field is reduced, thus lowering outward drift in the transverse direction. However, if it increases beyond a point, then it acts as a detriment to confinement by making the plasma unstable against kink-type perturbations. Thus, a reactor must maintain the right balance by having fewer particle revolutions around the minor as compared to the major axis so that ExB drifts and kink perturbations don't arise.

A flux surface, a surface on which magnetic field lines lie, is traced out by every field line as a result of its revolutions around the toroidal and poloidal axes. The motion of ions under such conditions is guided by a system of nested toroidal flux surfaces, created by the magnetic field lines. However, it is essential that the field lines don't cover the entire magnetic flux surface on which they lie, only recombining after some trips around the torus. This is due to the fact that having a rotational transform that is a multiple of 2π can give rise to undesired charge separation, akin to the effect of the grad-B force or curvature drift in a toroidal field. Any variation in the safety factor q from one magnetic flux surface to the next can cause a shear in the magnetic field (see fig 30). Because the magnetic field lines point in different directions, as a single one proceeds radially onto different flux surfaces, the value of q, in typical tokamak plasma, ranges from 1 near the centre to values ranging from 3-4 at the edge of the plasma.

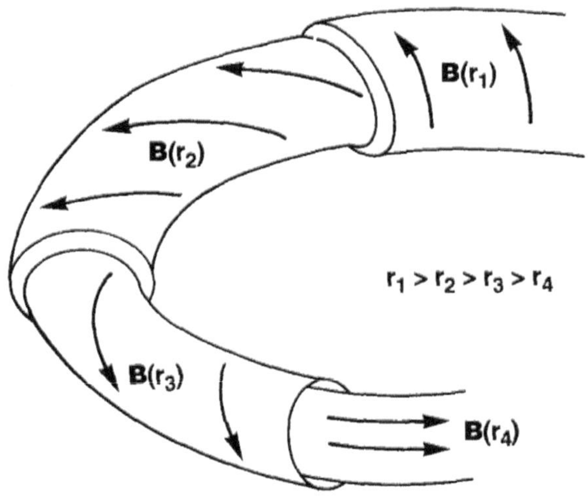


Figure 30: Nested magnetic flux surfaces with different shear

Magnetic shear as such, which results from the variation in the value of the safety factor, and hence, the rotational transform, is effective against the growth of both kink and drift perturbations. The way this works is by any perturbation aligned with any flux surface, say, B(r), with an increased minor radius distance r + dr encountering field lines at a different angle. Thus, the angle will vary as the perturbation grows to another distance r + dr', and all helically resonant instabilities will be radially localised. There exist two methods to establish a poloidal field in order to confine a plasma in a torus. The first involves the arrangement of a set of poloidal coils outside the plasma, while the second employs a toroidal current flowing in the plasma. The former is used in stellarators, while the latter is employed in tokamaks. A self-generated poloidal field is relatively simpler, however, and hence, tokamaks are of greater current interest. It is these devices which will now be considered.

## Tokamak

The tokamak, a Russian acronym that stands for "toroidal, kamera (chamber), magnet, and katuschka (coil), is the most widely known fusion device. Its components can briefly be summarised. First, there is the plasma torus, which acts as a single-winding secondary of a transformer. This allows for current to be induced by transformer action by means of the flow of current in the primary transformer winding. This current provides for ohmic heating while at the same time generating the poloidal magnetic field. The vector addition of the poloidal and toroidal fields is the basis for the rotational transform. It should be noted that these components share a relationship of perpendicularity, and their addition can be denoted by

$$\mathbf{B} = \mathbf{B}_{\phi}\ \mathbf{B}_{\theta}$$

In an attempt to understand the arrangement of poloidal and toroidal fields, as well as the components which are responsible for producing them, a diagram from *Principles of Fusion Energy: An Introduction to Fusion Energy for Students of Science and Engineering* by A. Harms, and Klaus F.. Schoepf has been included below.

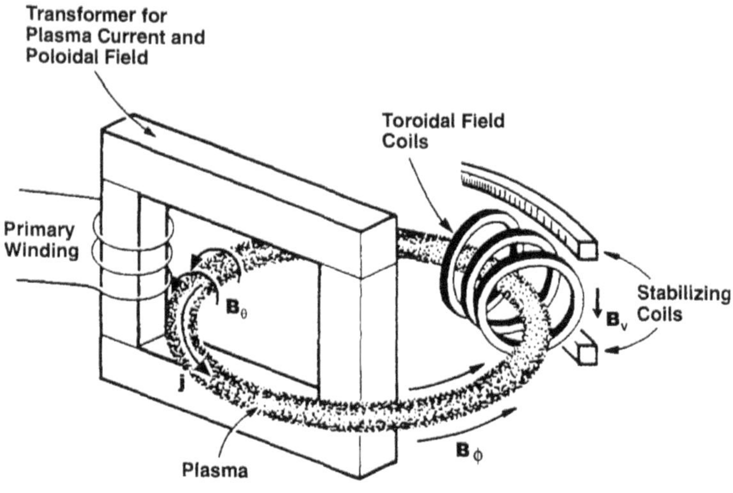


Figure 31: Components of a Tokamak

It is necessary to include stabilising coils in order to restrain the plasma from expanding, for which it becomes necessary to account for $B_v$ in the total field vector B. Here the difference between the poloidal magnetic pressure, $B_\theta^2 / 2\mu_0$, internally as compared to externally in terms of the torus, is responsible for the propensity of tokamak plasmas to expand. The larger poloidal field towards the centre steers the plasma ring towards assuming an increased major radius, thus necessitating the inclusion of a vertical magnetic field $B_v$ in order to confine the plasma. As indicated in the diagram above, the field is supplied by the stabilising coils. This field serves to prohibit the outwards expansion of the plasma by producing a 1 x $B_v$ force directed towards the centre of the reactor. The size of this field, however, depends on the toroidal plasma current, such that if the latter is sufficiently small, a smaller $B_v$ is needed. Hence, the magnetic confinement in a tokamak can be characterised as follows:

$$B_\phi > B_\theta > B_v$$

In such a toroidal device with a large plasma current that is used to supply poloidal magnetic induction and a powerful toroidal magnetic field, a toroidal electric field $E = (0, E_\phi, 0)$ is responsible for generating the former. This field is achieved by using transformers to temporarily change the magnetic flux $\psi_{trans}$ that penetrates the hole in the torus. The rate at which $\psi_{trans}$ is changed through the closed winding loop (hole in torus) is equal to the work done per unit charge by the induced electric field over the closed path along the winding of the secondary transformer (essentially the plasma ring). Iron or air core transformers are most commonly used to achieve such a flux change.

The induced toroidal magnetic field is given, applying Stoke's law, as

$$E_\phi = - (1 / 2\pi) (d\psi_{trans} / dt)$$

A consequence of the inductive scheme is the fact that such reactors are operated in a pulsed mode rather than a steady state. Generalising Ohm's law for a conducting fluid in a magnetic field gives the density of a plasma current driven by the induced toroidal electric field, with $\eta$ denoting the specific resistivity of the plasma and V representing the velocity of the fluid, as

$$\eta j = E + V \times B$$

Here the thermal plasma resistivity is based on the collisions within particles of the plasma and varies as

$$\eta \propto (kT_e)^{-3/2}$$

Evidently, this relationship of proportionality exists due to the fact that states of higher plasma temperature represent a decreased coulomb cross-section $\sigma_s$, and hence, a decreased rate at which particles dissipate their energy through collisions with each other. A response of the plasma to the induction of a plasma current suggests that ohmic dissipation- a loss of electric energy due to conversion into heat when the current flows through the plasma- is only viable up to temperatures below 1keV for reasons discussed in the previous chapter. Moreover, that this is the case is implied by the statement of proportionality above. Yet, in spite of the limitation of ohmic heating, the need for a toroidal current in the plasma remains due to the necessity of a poloidal field. The resistivity of the plasma is attributable to a larger extent to ions, as electrons have relatively fewer collisions, on the whole carrying the induced current. That this is the case is seen from the frequency of electron and ion collisions:

$$(\tau_{ei})^{-1} \propto \sigma_s v_r \propto v_r^{-3}$$

To illustrate, consider the case of electrons from the high energy tail of a velocity distribution, that is, those travelling with an above-average velocity. These electrons would move in the opposite direction to others, and as a result, exhibit a lower collision frequency. Moreover, this would allow them to accelerate, hence, increasing their velocity even further in an endless cycle. This phenomenon is known as electron runaway, and results in the existence of a portion of the electrons in a tokamak never colliding, forming an accelerated beam away from the main distribution. The only condition necessary for this to occur is a high electron velocity gain prompted by a rapid decline in the coulomb-cross section.

## Particle Trapping

Taking the case of a magnetic field which is axisymmetric, it is possible to make a few conclusions about the state of the reactor. Firstly, the flux surfaces embracing the field lines would approximate annular, or ring-shaped toroidal surfaces with r = constant. Moreover, the charged particles in such a scheme would follow helical field lines that result from the combination of the action of the toroidal and poloidal fields. Finally, a natural corollary of these two is the fact that the particles would move on the flux surfaces with the exception of excursion of the order of their gyroradius, ie. when the centre around which particles gyrate is close enough to the end of the surface such that they periodically exit said surface following a helical trajectory. With these assumptions, it is possible that some particles may remain trapped in magnetic wells, unable to spiral around helical field lines while encircling the major and minor torus axes. This well is formed by the variation in the strength of the magnetic field between the inboard and outboard sides of the torus. Generally, the strength of both the poloidal and toroidal fields tends to be greater towards the centre, resulting in an overall field variation as shown in fig. 32. In consequence, the propensity for particles to move on flux surfaces, combined with the existence of a gradient in the strength of the magnetic field, results essentially amounts to the existence of a set of magnetic mirrors.

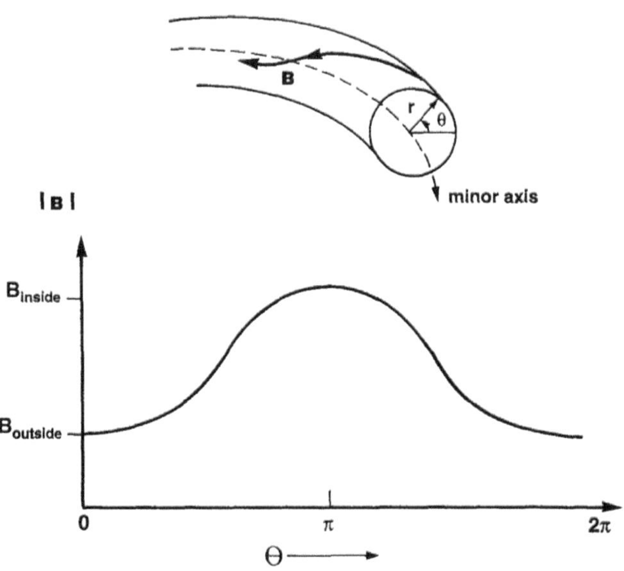


Figure 32: Poloidal variation of the overall magnetic field in a tokamak

Recall that mirrors trap particles with a low parallel velocity component compared to the so-called "passing particles"- those moving fully around the torus. Viewed from a transverse plane of the torus, the particle trajectory from a number of toroidal journeys shows that particles encounter mirror reflection in banana-shaped orbits as such:

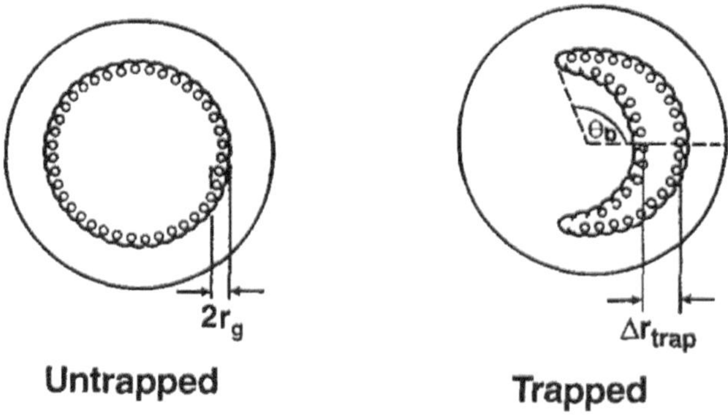


Figure 33: Projection of particle orbits onto a poloidal plane

Here $\Delta r_{trap}$ indicates the banana width of the orbit of the trapped particle. It is possible to write an expression for the fraction of particles trapped in the mirror field as was done earlier in the chapter, as well as with the assumption that B < < B so that $B \approx B_\Phi \propto 1/R_0 + r \cos \theta$, respectively as

$$f_{trap} \equiv \sqrt{1 - (B_{min} / B_{max})}$$

and

$$f_{trap} = \sqrt{1-[(1-\varepsilon) / (1+\varepsilon)]} = \sqrt{2\varepsilon / 1+\varepsilon}$$

Here $\varepsilon = a / R_0$ = inverse aspect ratio of a tokamak. Thus, for a tokamak for which $a / R_o \approx ⅓$, approximately 70% of plasma particles appear to be trapped. Additionally, enhanced particle leakage can be associated with banana trapping. The maximum distance which trapped particles can be displaced is as great as the width of the banana $\Delta r_{trap}$, while for an untrapped particle, it is of the order of gyration radius rg. It is possible for the former group of particles to escape from the tokamak faster than those untrapped, as typically $\Delta r_{trap}$>rg as well as the fact that D (coefficient for diffusion perpendicular to the magnetic field) increases with $r_g^2 = v^{-2} /a)_g^2$ and $\Delta r_{trap}$ respectively. Consequently, rather counterintuitively, having a larger fraction of trapped particles constitutes a substantial problem in tokamak confinement due to the large leakage enhancement of the sort described above.

## Tokamak Equilibrium

Like other magnetic confinement concepts, tokamaks require a state of equilibrium in order to prevent the plasma from making contact with the walls of the vessel. In this context, it is useful to view this fusion plasma as a single fluid subject to a magnetohydrodynamic description. Thus, it is necessary for the state of equilibrium to be subject to MHD equations for the plasma current density j and magnetic field B, as well as be in accord with Maxwell's equations. Specifically, the equation of interest in a magnetostatic field (a system where the currents are steady-not changing with time) is:

$$\nabla \times \mathbf{B} = \mu_0 \mathbf{j}$$

$$\nabla \cdot \mathbf{B} = 0$$

Therefore, the equation for balance can be given as

$$\nabla \mathbf{p} = \mathbf{j} \times \mathbf{B}$$

This implies that both the electric and magnetic fields lie on an isobaric surface. Further, it is possible to determine the following equation and its vectorial form respectively by combining the equations of the strength of the poloidal magnetic field strength and the flux surface function ψ as

$$\mathbf{RB_\theta = d\psi / d'r}$$

and

$$\mathbf{B_0 = -(1 / R)\, e_\Phi \times \nabla\psi}$$

At this point it would be useful to visualise the action of the electric and magnetic field within the framework of a cylindrical coordinate system to better be able to describe their individual components (see fig. 34). Due to the symmetry of the electric and magnetic field in the force balance equation, the poloidal current function F(ψ) exists. The poloidal component of current density can here be expressed as

$$\mathbf{J_\theta = -(1/R)\, e_\Phi \times \nabla F}$$

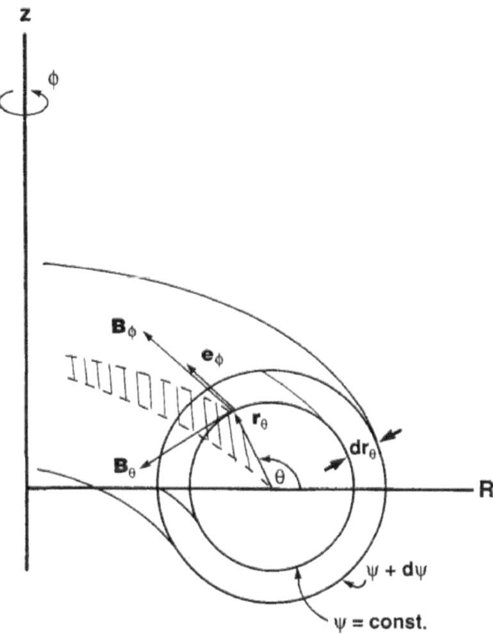


Figure 34: Magnetic field components in a cylindrical coordinate system

Note that here ψ, F, and $B_\Phi$ x R remain constant on the magnetic surface. Moreover, with Maxwell's equations the toroidal field can be defined within this framework as

$$\mathbf{B_\Phi = \mu_0 (F / R) e_\Phi}$$

In addition, it is possible to decompose the magnetic field into its constituent components along the R and Z axes, yielding the following:

$$\mathbf{B_R = -(1/R) (\partial\psi / \partial z)}$$

and

$$\mathbf{B_z = (1/R) (\partial\psi / \partial\psi)}$$

The reconciliation of all of the above equations, the introduction of flux and current function in the force balance equation yields the famous Grad-shafranov equation as

$$\Delta^* \psi = (\partial^2\psi/\partial R^2) - (1/R)\,(\partial\psi/\partial\psi\,) + (\partial^2\psi/\partial z^2) = -\mu_0 R^2\,(dp/d\psi)\ -F(\psi)\,(dF/d\psi)$$

The poloidal magnetic field induced by transformer action provides a few boundary conditions. For a state of equilibrium, the equation above shows that the magnetic field requires a uniform vertical part $B_v$ to the radial expansion of the current-carrying plasma. The manner in which the topological structure of the magnetic field is altered with the imposition of $B_v$ is depicted in fig. 35, with the plasma current j that induced the initial poloidal field flowing into the plane of the page. A numerical solution of the Grad-shafranov equation would give not only the geometrical location of the magnetic surfaces but the radial distribution of the axial current density as well. This distribution has been found to be consistent with experimentally measured pressure profiles and an externally applied magnetic field. Note that any closed nested flux surfaces inside the separatrix are the result of solutions of the equation when the plasma pressures are not very high.

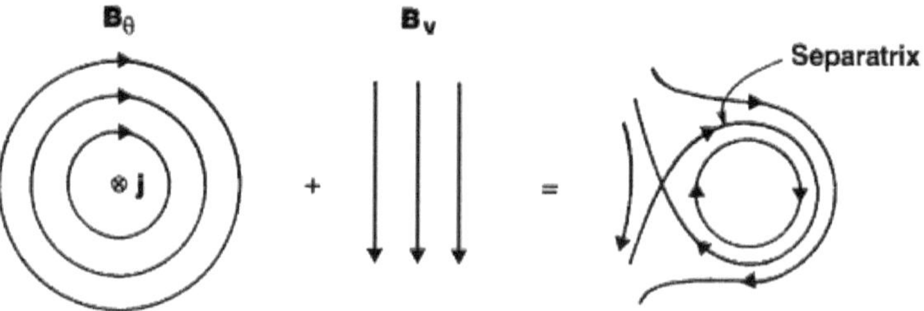


Figure 35: Superimposing a vertical magnetic field onto an induced poloidal field

In a tokamak plasma it becomes apparent that there exist various equilibria, all of which are characterised by the ratio of the plasma pressure averaged over its volume to the average energy density of a poloidal magnetic field over the magnetic surfaces at r=a. This so-called poloidal beta can be expressed as such:

$$\beta_p = \langle p \rangle / B_\theta^2 (a) / 2\mu_0$$

And even though the total beta of a tokamak is constrained to values of a few per cent as a result of its inability to achieve magnetohydrodynamic stability, the poloidal beta can tend to appear between 0.1 and 2.5 for the regimes of equilibrium displayed in figure_, with $\ell_i$ denoting the plasma internal inductance per unit length. The role of the toroidal field is that it contributes to the balance of pressure, in a manner affected by components of the poloidal current, by a difference given by:

$$\langle B_\Phi^2 \rangle_{\psi(a)} - \langle B_\Phi^2 \rangle$$

These components may either increase or decrease the presence of the toroidal field in the plasma. Specifically, if the square of the strength of the toroidal field averaged over the volume of the reactor is greater than that at the average of the outer flux surface, then the current is called paramagnetic. If, however, it is the opposite case, and the square of the field strength is greater, on average, at the outer flux surface, then the current is called diamagnetic.

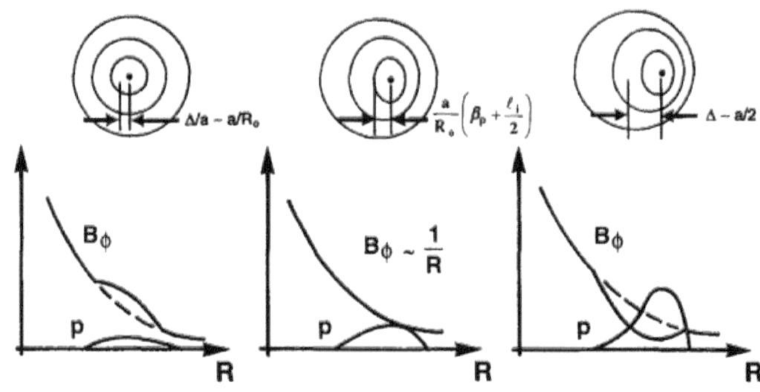


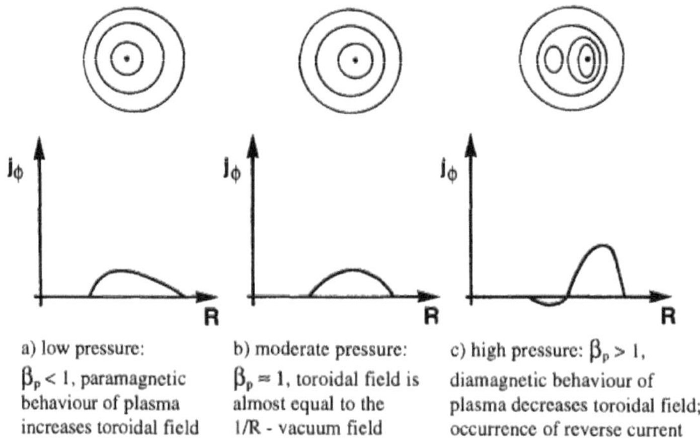


Figure 36: Equilibrium characteristics varying with plasma pressure

Thus, a diamagnetic plasma is associated with a plasma pressure greater than $\langle B_\Phi^2\rangle_{\psi(a}/2\mu_0$. That is, with a poloidal beta value greater than 1. In addition, the magnetic axis of a column of the torus is displaced outwards with increasing plasma pressure until the displacement is roughly equal to half of the minor plasma radius. This phenomenon represents the limit of a state of equilibrium. Moreover, the poloidal magnetic field must support at least a fraction $a/R_0$ of the kinetic pressure such that:

$$B_\theta^2 (a) / 2\mu_0 > (a / R_0) <p>$$

This means that the poloidal beta is restricted to values less than $R_0 / a$. Further, the introduction of the Kruskal-Shafranov limit, as well as the assumption that $B_\theta^2$ (a) and $B_\Phi^2$ are approximately equal to $\langle B_\theta^2 \rangle_{\psi(a)}$ and $\langle B_\Phi^2 \rangle_{\psi(a)}$ respectively, yields an expression of $B_\theta^2$ (a) and $B_\Phi^2$ by means of their beta value to obtain the following relation:

$$<p> / (B_\Phi^2 / 2\mu_0): = \beta_t \approx (a^2 / R_0^2) (\beta_0 / q^2(a)) \leq (1 / 6.25) (a / p_0)$$

Neglecting the inferior magnetic pressure of the vertical field $B_v$, the total plasma beta can be given, and subsequently constrained respectively, by

$$\beta = <p> / (B_0^2 (a) / 2\mu_0) + (B_\Phi^2 / 2\mu_0)$$

and

$$\beta < (a / R_0) / (q^2(a) + (a^2 / R_0^2) < (a / R_0) / 6.25 + (a^2 / R_0^2)$$

As the aspect ratio of a particular reactor increases, this value proceeds to essentially become the toroidal beta $\beta_t$. Hence, a tokamak with an aspect ratio, that is, the ratio of its major to minor axis, approximating 3 will be limited to a beta value below 5% as per the magnetohydrodynamic stability considerations. In this case, the poloidal beta may amount to a value in the order of $R_0 / a$. A consequence of the limitation on the poloidal beta is the constraint:

$$I < (2\pi / \mu_0) [a^2 / R_0 q(a)] B_\Phi]$$

This comes, by virtue of Ampere's law, from the expression of the poloidal magnetic field induction by the plasma current generating the field. Evidently, it signifies that there exists a limit beyond which the plasma current cannot be increased in order to achieve higher temperatures with ohmic heating. The magnitude of the plasma current is bound by the need for stability, and hence, constantly increasing the plasma current arbitrarily is unviable. However, the above derivations of a set of constraints for the plasma beta are rough, and hence, the subsequent section will more definitively expound the stability of tokamaks. Thus far the discussions of the variation of the toroidal and poloidal magnetic induction vectors with both toroidal and poloidal rotation can be summarised using the notion employed above as follows:

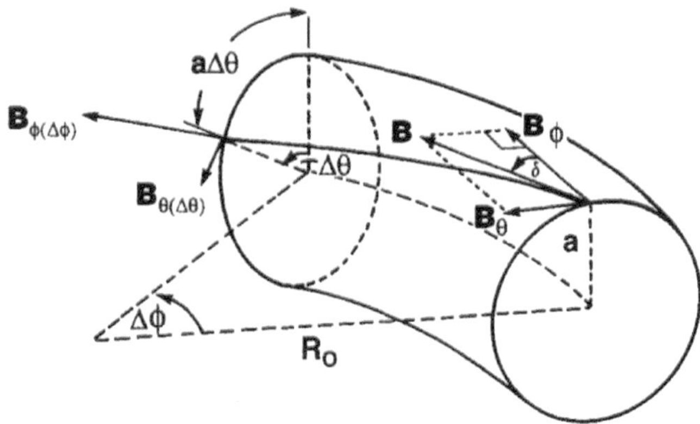


Figure 37: Induction vectors varying with toroidal and poloidal rotation

## Stability of Tokamaks

It should principally be noted that a state of equilibrium that doesn't display an exact Maxwellian distribution, and hence, is not in perfect thermodynamic equilibrium, is inherently unstable. This is due to the fact that even though the balance of forces is zero, and a stationary solution exists to the relevant equations, the entropy within the system has not reached the maximum level possible. That is, free energy appears available, which serves to excite and give rise to perturbations. This free energy is associated with the currents which eventually align themselves parallel to the magnetic field, thus making a force-free current (the Lorentz force is equal to zero), as well as the plasma pressure. These factors determine the stability of a plasma confined by toroidal and magnetic field configurations, and hence, they will here be considered.

First, it is important to establish the relevant factors that constitute an MHD mode. The ratio of the two forms of free energy mentioned above can be denoted by $\beta_p$. Recall the 'bad' magnetic field curvature considered for open confinement systems, which can be attributed fundamentally to two destabilising forces: the gradient of both the magnitude of the plasma current ($\nabla j$) and the pressure ($\nabla p$). These forces inherent to tokamaks can be divided into ideal MHD modes, resistive MHD modes, and microinstabilities, each of which requires individual consideration.

### Ideal MHD Modes

Destabilising forces of this sort can be the most virulent due to the fast rate at which they grow, coupled with their ability to extend over the entire plasma rather than remaining localised. The combination of these two factors results in the resistivity of the plasma within the short time scale of relevance (a few microseconds) being negligible. Here the bad convex curvature is present in the helical magnetic field lines on the outboard side of the torus, thus making such a scheme subject to flute-type interchange instabilities. In order for the curvature of the magnetic field lines over a complete poloidal rotation to be favourable, the windings must have a rotational transform that is below 2π, ie. with a safety factor equal to or greater than one. Interestingly, interchange perturbations don't grow in tokamaks which exhibit a value of q equal to or greater than one, as, seen from the plasma, the inboard side exhibits a concave curvature of the helical field lines- those concentrated on this side maintain a 'good' curvature in a tokamak. Nevertheless, these perturbations can grow locally in the outboard region due to their unfavourable curvature. Such is known as 'ballooning', and hence, is referred to as the ballooning instability. Recall that such instabilities are a result of a gradient in the plasma current magnitude and pressure, although in this case, only the latter is responsible for driving the instability. Evidently, then, the solution to a high-pressure gradient is necessary to suppress such modes described, for which there exist two options that are applicable in almost any region of the reactor- the establishment of appropriate pressure profiles and of the appropriate magnetic field line windings which correspond to a favourable curvature.

In such a mode the kink instability has been found to be the most dramatic, seeing as it causes a contortion of the helical plasma column, and hence, of the magnetic flux surfaces that it comprises. This instability has a proclivity to occur in tokamak plasmas which have a low pressure, and is driven by the radial gradient of the toroidal current. Specifically, these instabilities are bounded to small intervals of q(a) that lie close below integer values. When this value is less than two, for the typical current profile distributions for tokamak plasmas, modes which are unstable arise. The associated kink distortion of the plasma column can be stabilised by the enhancement of the strength of the toroidal magnetic field in such a manner that the Kruskal Shafranov condition, which can be expressed as follows, is fulfilled

$$B_\theta / B_\Phi < a / R_0$$

A part of the reason why the plasma beta can't assume just any value is the fact that the safety factor needs to be adjusted to the appropriate value for the plasma to remain stable against the kink distortion mentioned above. The maximum limit for this value, then, can be expressed as a critical beta as such:

$$\beta_{crit} [\%] = C_T (I/aB)$$

Here $C_T$ represents the Troyon factor, a dimensionless value, that ranges from 2.8 to 5, depending on the nature of the instability at hand. If the plasma current I is in MA, then the minor radius a is in metres and the confining field in Teslas. It has been found necessary to include the Troyon factor in order to account for the actual cross-sectional contour for a non-circular shape of the plasma cross-section. Evidently, the shape of a tokamak cross section is not a circle, but rather elliptic, bean, or D-shaped. This is to permit the optimisation of energy confinement as well as the desired plasma pressure profiles.

### Resistive MHD Modes

Other instabilities which occur in tokamak plasmas can be attributable to the electrical resistivity within them. This makes instabilities grow at a slower rate, yet can be detrimental to confinement by destroying the nested topology of the magnetic flux surfaces that is necessary to localise perturbational effects. This mode is a result of the diffusion or tearing of the magnetic field lines relative to the plasma fluid. Such diffusion may preponderate the ideal MHD effects for helically resonant magnetic field perturbations. The manner in which this occurs is that, in the thin boundary layers around surfaces which posses a rational value for their safety factor, the magnetic field lines re-arrange themselves in a manner that produces non-axisymmetric helical islands as suggested by the figure below:

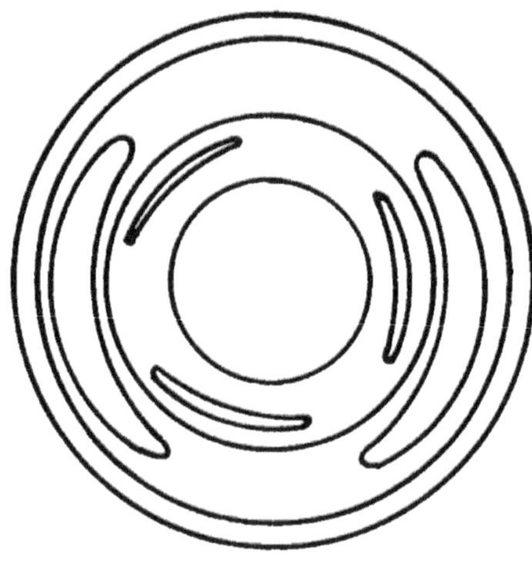

Figure 38: Cross-sectional view of magnetic islands at flux surfaces

This instability is driven by the radial gradient of the plasma current density present in the tokamak. Moreover, growth occurs upon the formation of a magnetic island filament- a process that continues until all the free energy accessible from the current is acquired by the instability. At this point, the plasma undergoes a sequence of MHD equilibria outside the so-called resonant surfaces- those on which q assumes a rational number. Thus, in plasmas confined with a weaker magnetic field, it is possible for such resistive modes to couple non-linearly amongst each other, leading to the complete disruption of the current of the plasma. However, the increase of this current makes it progressively more stable against tearing modes, as it makes them diminished. Increasing the current in such a manner is associated with lowering the value of the safety factor q.

Another effect of the non-linear evolution of resistive MHD modes has been observed. That is, certain plasma parameters have exhibited so-called "sawtooth" behaviour. These include the density of the electrons at the centre of the plasma column (where the value of q can be less than 1), as well as the temperature of the electrons. Natheless, Sawtooth oscillations can be prevented by modifying the current profile near the magnetic surfaces on which q=1. This is because such oscillations are constrained within the internal regions of such surfaces.

### Microinstabilities

This mode is typically found in velocity distributions which are non-Maxwellian. The manner in which instabilities occur within this mode will now be summarised. Firstly, deviations from a state of thermodynamic equilibrium result in the presence of free energy. This free energy drives instabilities, which eventually evolve into turbulences that engulf the entire plasma. In addition, other velocity distributions which can give rise to instabilities include those which are non-uniform and anisotropic (where the velocities vary with the direction in which the particles are travelling). In the case of both of these distributions, the particles in a plasma can assume, it is necessary to employ a kinetic, rather than a fluid description in order to account for the effects of particle kinetic energy that cause instabilities in the first place. The velocity distribution tends towards anisotropy as the density of the plasma decreases, as it is necessary for individual electrons to align their velocities in the direction of the current density in order to carry a given plasma current. This drift velocity has the proclivity to make a velocity distribution function more and more asymmetric, and hence, unstable. On the other hand, the velocity distribution can be established by the collisions which electrons in the plasma encounter. This is due to the fact that such collisions randomise their velocities. This stabilising effect is ultimately overcome, however, as the density is further decreased, which, as mentioned earlier, due to its correlation with an anisotropic electron velocity distribution, results in a drift velocity that destabilises the plasma.

On the whole, any method by which plasma is confined by mirror fields is characterised by its propensity to have its electrons assume an anisotropic velocity distribution due to the fact that particles with a large parallel velocity component will escape, and therefore are lost from the distribution. Hence, the particles trapped in a tokamak which bounce back and forth in the local mirror fields can constitute a significant source of instabilities. Moreover, they can do so to a greater extent when the frequency of perturbations is exceeded by the frequency at which the particles are bouncing in such a manner. The instabilities associated with such a state are collectively referred to as trapped particle instabilities and currently remain in need of further investigation for their effects on the increase in cross-field diffusion in tokamaks.

On the topic of drifts, it is important to make the distinction between different species of particles on the basis of their motions relative to one another. In this regard, for the sake of simplicity, this relative drift will be considered both when current is driven through the plasma and when a beam of highly energetic particles is injected. Firstly, the waves in a plasma are excitable by the energy of drifts. It is possible for a disturbance to grow by the transfer of drift energy to oscillation energy. The resulting instabilities are called the ‘two stream instability’ and the 'beam-plasma instability'. Secondly, it is possible for an instability to occur under the existence of a steep density gradient in the plasma. Known as the 'drift instability', it is caused by electron drift and hence, can only be stabilised by the addition of appropriate magnetic shear.

In conclusion, micro-instabilities constitute a large and complex field in plasma physics, still under investigation and many of whose theoretical predictions yet remain to be proven experimentally. Generally, however, three effective methods to prevent plasma instabilities have been found. These are the introduction of a magnetic shear, the operation of a minimum-B configuration, and the dynamic stabilisation by the oscillation of the electric and magnetic fields involved in the confinement of the plasma. In addition, the achievement of dynamic stability by the proper-phase force feedback can also be considered as a part of the third method. In all, while the description of a tokamak in this section has remained brief, the reader is provided with a rudimentary understanding of the scientific principles, components, as well as the potential and actual shortcomings of tokamaks so as to enable the comparison of the concept to other closed magnetic systems that are to be discussed.

## Stellarator

A condition for confinement by means of a toroidal field is for the field lines to have a rotational transform. This is to prevent the plasma from assuming local charge concentrations or polarisation, as well as drifting towards the outer wall. The stellarator concept features a similar geometry to a tokamak but differs in that it renders a magnetic field with helical coils. This means that the device compels the field lines to have a rotational transform, by the helical or contorted coil current external to the plasma, or through the deformation of the torus itself. Early concepts that attempted to do this demonstrated poor stability, while recent attempts have been able to contort the magnetic field lines into a spiral form by passing currents through helical conductors wound around the torus. In this way, closed magnetic flux surfaces and a rotational transform are introduced to the plasma. Thus, stellarators don't rely on a pulse-induced plasma current, and all magnetic fields employed to confine the plasma are generated by the flow of current in external conductors. The figure below demonstrates how alternate coils produce helical currents in opposite directions.

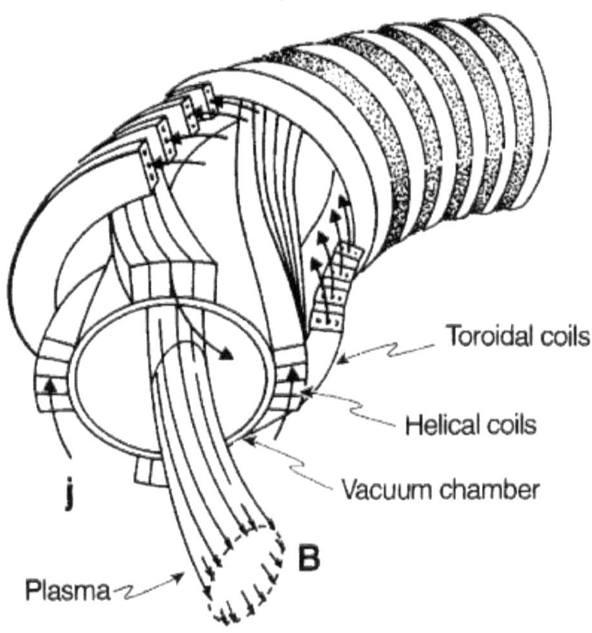


Figure 39: Magnetic field generated by external helical currents in alternate coils

Such a scheme allows for the plasma to be confined in a steady state, with the reactor operating continuously. The helical winding, which acts as a loosely wrapped solenoidal winding, generates the toroidal and poloidal field. In addition, it creates a vertical field akin to the one formed at the end of the plasma boundary in tokamaks. Hence, the currents in adjacent helical windings of the same pitch flow in opposite directions. On average, the effect of this is the cancellation of each other's toroidal and vertical fields. Thus, it is essential to introduce a separate set of coils to produce the essential toroidal magnetic field. Flux in the reactor results from the poloidal field, generated by the helical windings, together with the toroidal field, produced by additional field coils. This flux twists the magnetic field lines during their passage along the torus, thus, generating a magnetic surface with the following shape:

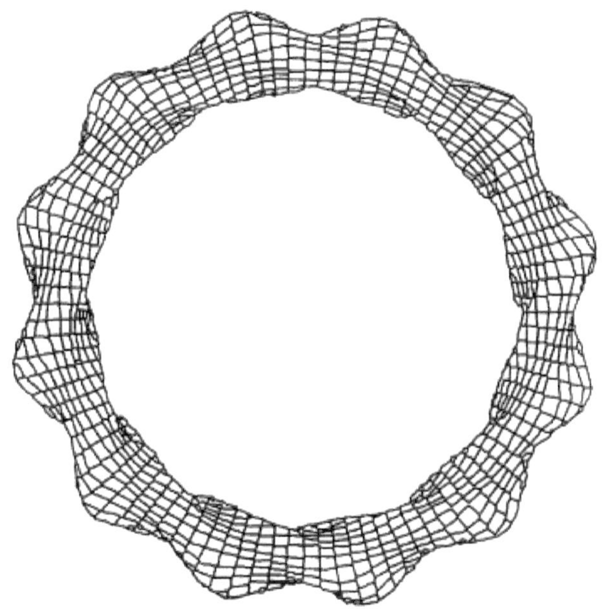

Figure 40: Complete magnetic surface viewed from the top of the stellarator

The geometric simplicity of a state of axisymmetry is now lost. Interestingly, closed magnetic surfaces necessitate a restricted mirror cross-section region of the toroidal tube, and it is only possible to establish a closed magnetic surface within this region. For a stellarator with $\ell$ = 3 pairs of helical coils with current in opposite directions, the shape of the generated magnetic surfaces can be depicted in the cross-section of the nested magnetic surfaces as such

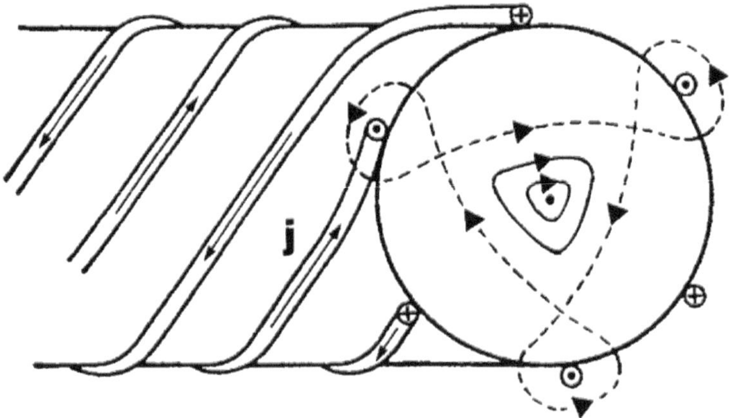


Figure 41: Cross section of nested magnetic surfaces generated by an $\ell$ = 3 stellarator

Closed magnetic surfaces in such a scheme occur at the cross-sectional area embraced by the dashed separatrix line, and hence, this area can be identified as the last closed magnetic surface. Beyond the separatrix, however, the field lines wrap around the individual conductors. And, because there is no flow of current in a stellarator, the line integral of the poloidal magnetic field vanishes along a contour encircling the magnetic axis on each flux surface. This means that along the contour, the poloidal field must change both its sign and magnitude. Hence, the magnetic field lines are not monotonically wrapped around the toroidal tube, but, unlike a tokamak, oscillate periodically. This oscillation occurs according to the qualitative structure provided by part a of the figure below.

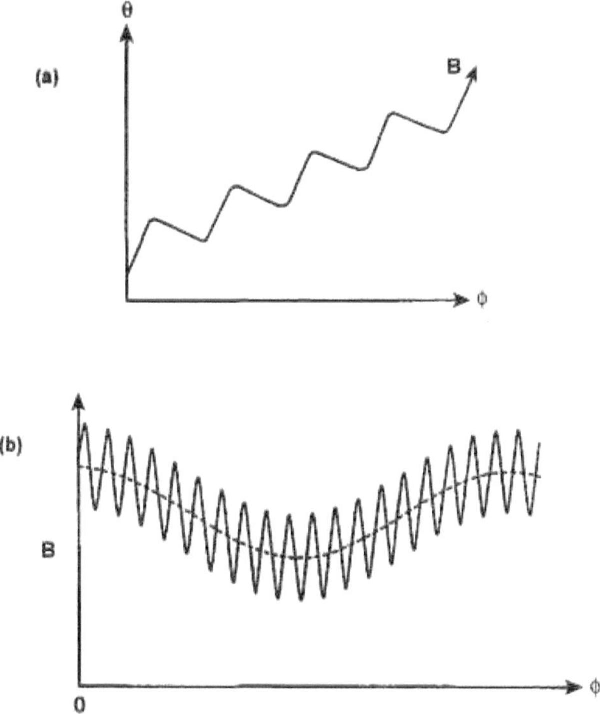


Figure 42: (a) Oscillatory behaviour of a magnetic field line on a flux surface. (b) The varying magnitude of the magnetic field

It is important to note that these field lines are incrementally rotated in the poloidal direction. However, the curvature associated with such a torus geometry results in an inhomogeneity of the magnetic field. Figure b above highlights this variation of the magnetic field line along the toroidal direction. Furthermore, that alternative coils carry current in different directions increases the frequency of oscillations in the magnetic field. The speed at which the modulation of the field occurs in such conditions corresponds to the toroidal curvature- The greater the curvature the slower the modulation. Evidently, the externally imposed helical field exposes the stellarator concept to numerous complications, and hence, appears detrimental to its ability to confine plasma viably. However, before such a conclusion can be reached, it is necessary to consider the characteristics of helical fields that make them desirable for closed magnetic confinement.

First, it is important to illustrate the fact that toroidal fields establish magnetic mirrors in a manner described earlier in the chapter, which traps the particles of the plasma. Helical fields do the same, although through the generation of local mirrors. The particles trapped in such schemes principally orbit around the centre of the torus in three different ways. They may simply be circulating particles, those which pass entirely around the torus without any encounter that results in a reflection. Next, there may be particles which are trapped helically, and hence, reflected in the local mirrors of the helical field. And finally, there are those that are trapped toroidally, that is, tracing banana orbits as a result of being reflected in the magnetic mirrors of toroidal fields. Note that it is also possible for this second group of particles to exhibit the behaviour of those in the third, also known as 'superbanana particles', depending on the individual specifications of the helical field under whose influence they are moving. Thus far, the method by which non-self-generated magnetic field confinement systems work has involved alternating coils on the basis of the direction of the flow of current through them. There is, however, another method by which a rotational transform can be introduced in the plasma to keep it from escaping confinement- the partial rotation of non-circular toroidal field coils relative to each other.

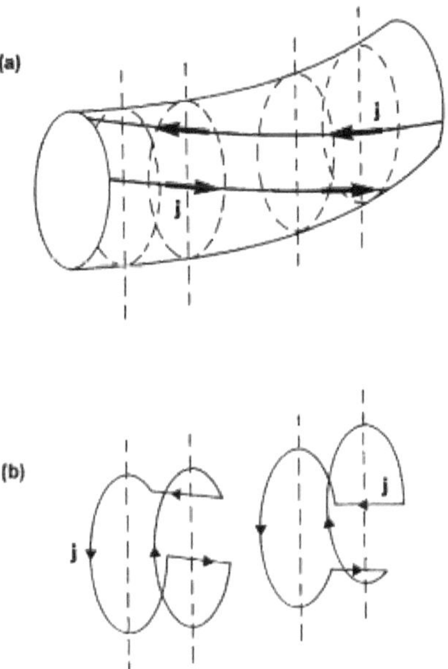


Figure 43: Depiction of replacement of helical conductors by modular elements

The figure above illustrates the fact that specific designs of stellarators can allow for the modular composition of their components. Part a shows the effect of helical stellarator windings as described above, while b displays modular coils which are generating similar conditions. The ways in which modular designs can offer solutions to the challenges faced by stellarators will be considered. For the sake of simplicity, the example of a stellarator consisting of two conductors with currents with opposite signs will be considered. This configuration would allow for it to be replaced by modules which combine parts of the helical coils while including additional conductors along the meridian. If more advanced modular designs are considered, however, then those which employ non-planar twisted coils appear to be the most feasible. This includes the Wendelstein VII-X, currently the most prominent reactor to employ the stellarator concept, whose coils comprise sophisticated spatial elements as such

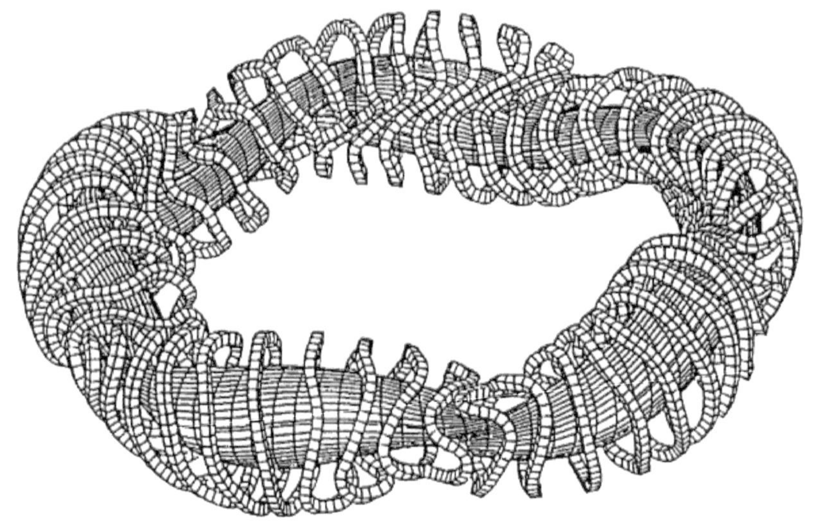

Figure 44: Design of the advanced modular coil stellarator Wendelstein VII-X

## Alternate Closed Configurations

Finally, there exist other closed magnetic confinement systems which differ from tokamaks and stellarators enough to warrant their own section. The aims of these reactors in terms of energy production are identical to those considered above, yet they differ significantly in terms of geometry, size, input power requirements, time scales, and technology. Note that while tokamaks and stellarators are believed to be the most viable form of plasma confinement, in the search for a new power source, no stone should be left unturned. That is, it is possible for there to be multiple concepts to yield more energy than they require, and there is certainly something to be learned about their behaviour if they do some. However, even the concepts which fail reveal some insights into the nature of plasma that can be useful for the construction of fusion devices. In all, the exploration of new concepts in fusion has always proved to be a worthwhile endeavour, and it would be impossible to cover the various magnetic confinement systems without addressing the schemes discussed here. Finally, the list of nuclear fusion startups doesn't seem to stop growing, with each attempting a novel method of achieving not only break-even but ignition as well.

## Bumpy Torus

The first of such configurations is the bumpy torus. This concept involves linking the ends of multiple magnetic mirror sections to form a high-aspect-ratio torus. The coils used to create these sections are equally spaced in the mirror array, and hence, equidistant from each other. The plasma threads the bores of the axisymmetric coils, and hence, assumed a bumpy toroidal 'sausage' shape, with the plasma column squeezed together near the regions of $B_{Max}$ and occupying the largest volume at the regions of $B_{Min}$ (areas furthest from the coils). In such a regime, the magnetic field lines close in on themselves. The features here described can be depicted by the following

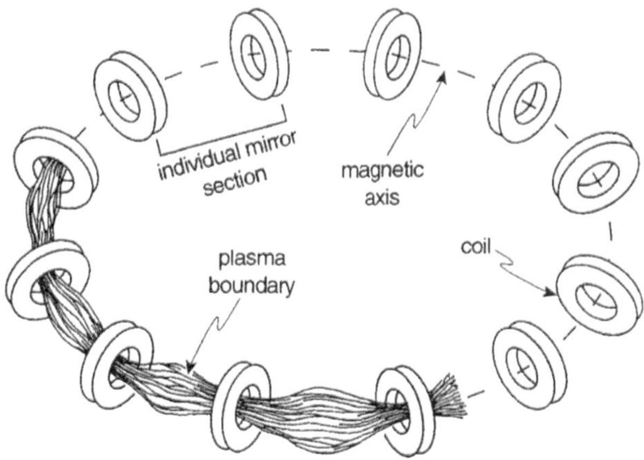


Figure 45: Torus comprised of linked magnetic mirror sections

Particles in a bumpy torus are confined in two ways. First, particles are trapped and reflect back and forth within individual mirror sections, the manner in which they do so described in the open magnetic confinement section. Second, particles circulate around the major circumference of the plasma of the bumpy torus. In order to stabilise the configuration as well as heat the plasma, electron cyclotron resonance heating (ECRH), discussed previously in Chapter 1, is applied, forming an annular, high-energy electron plasma in the central part of the torus. The viability of this as a heating method is evident, while the method by which the configuration is made stable by the currents generated by the hot-electron rings which provide a minimum-B configuration. A reactor that does this is called the ELMO (Electrons with Large Magnetic Orbits) bumpy torus. The temperatures which the hot-electron annuli would have in most cases is below 100 keV, while their density is approximately $10^{18}m^{-3}$. This makes the temperature and density of the plasma 15 keV and in the order of $10^{20}m_{-3}$ respectively. Thus, the beta value of the plasma confined in a bumpy torus can be comparable to that of the annuli alone.

A condition for macroscopic stability is for the toroidal beta value to be roughly equal to or greater than the annuli beta, which, if it is to provide a minimum-B configuration, must surpass a certain threshold that varies between 5-15% based on the shape of the annuli. These requirements significantly hinder the viability of bumpy torus schemes, although the production of annuli betas that comprise 50% of the total beta in a steady state has been demonstrated experimentally. Hence, it is possible for the core plasma beta of the ELMO reactor to be established at, relative to tokamaks, substantially increased values. A consequence of this is the fact that the fusion power density, as limited by the magnetic pressure, is elevated. Or, for the given fusion power density, the magnetic field requirements are reduced significantly.

There are some significant advantages of a bumpy torus over tokamaks. Mainly, the fact that they have a larger aspect ratio (around 5-10 times larger) makes their design and construction on the whole far easier. Furthermore, they can be operated in a steady-state mode as a consequence of not requiring any form of power interruption. A major limitation of concepts like ELMO is the fact that it is not necessary that the central hot-electron annuli are produced due to the relative inefficiency of ECRH in this regard, as well as the resulting radiation losses that would make sustaining such a state unviable. However, it is possible to address the need for a minimum-B configuration by toroidally linking the individual modular coils. The shape of the coils is such that each already represents a minimum-B magnetic mirror. A rotational transform is introduced by the rotation of these coils, which otherwise don't exhibit any form of poloidal symmetry, around a magnetic axis with respect to each adjacent coil.

### Screw Pinch

Consider a device combining the effect of a z-pinch and θ-pinch. Such a device would contain a plasma with currents in both axial and poloidal lines. Thus, it would generate a magnetic field that confines the plasma with helical field lines. This concept is called a screw pinch due to the form assumed by the field lines. Akin to a tokamak, it is operated at a relatively high beta (≈ 20%), although it only features plasma confinement to a sufficient degree for very short periods.

## Reversed Field Pinch

Another concept that demands toroidal confinement is known as the reversed field pinch (RFP). It has received great attention due to the improved stability gains and favourable magnetohydrodynamic modes it exhibits. Akin to a tokamak, it involves confining the plasma by a combination of toroidal and poloidal fields. In addition, the poloidal field is generated by the toroidal plasma current by transformer action, while the toroidal field is primarily established by the external coils. There exists, however, an essential difference. In reversed field pinches, the plasma currents are parallel to the magnetic toroidal minor axis and alter the toroidal field in a diamagnetic fashion. This alteration is such that the toroidal field can essentially change its sign near the plasma boundary. This phenomenon is known as field reversal and is suggested to emerge from a turbulent state as a method of self-organisation. On top of that, the plasma current and poloidal field are stronger than a comparable tokamak, whereas the toroidal field is comparatively modest. A consequence of the disparity in field strength results in strongly sheared magnetic field lines, whose pitch increases rapidly with greater radial distance. The reversed field pinch configuration is produced by the high magnetic shear near the edge of the plasma, which is responsible for suppressing local MHD instabilities.

Unlike a tokamak, which has the safety factor q acting as a condition for viability, these configurations need to meet the kruskal-shafranov stability criterion- $q(r) > 1$ everywhere and $q(r=a) \geq 2.5$. Current RFP configurations feature a value below 1 for the former constraint and a negative value for the latter, consistent with the reversal of the toroidal field component in the edge region. The transition from a tokamak to an RFP plasma requires the presence of electricity- a conducting shell just outside the toroidal plasma or the presence of closely fitting external conductors. Both are essential to assist a tokamak plasma to remain confined while reducing q and turning into a reversed field pinch configuration. According to the theory of magnetohydrodynamic stability for reversed field pinch devices, a limitation for the plasma beta is somewhere around 30%. Seeing as this value is high, it is possible for these devices to be operated in advanced fusion fuel cycles. If the system wasn't constrained by the kruskal-shafranov stability criterion, then it would be possible to heat it ohmically to the point of ignition so long as the energy confinement is robust. In all, while both tokamaks and reversed field pinch devices are plagued by the need to operate in a pulsed mode, that tokamaks require their aspect ratio to be minimised, as well as the simpler design and maintenance of reversed field configurations, suggests that RFP devices have a sporting chance to offer the promise of clean energy in the near future.

## Spheromak

Finally, while all magnetic confinement systems so far aim to simply achieve the highest possible energy yield, they fail to consider the practical challenges associated with the reliance on fusion energy as a power source. For instance, all devices so far discussed are physically large, employ complex technology, are expensive to build, and ultimately possess a relatively low power density. The last point is of particular importance, and any concept must be underpinned by the need to address it. To that end, there exist concepts which achieve the same total power of reactors considered thus far, only in a significantly smaller geometry. Examples of such devices include compact RFP configurations and spheromaks. And as the underlying principles of the former have been discussed above, it is the latter that will now be considered. Spheromak configurations are characterised by an extremely low ratio and the absence of an external toroidal field. The structure of the device appears like a sphere, with a minor plasma radius that approximates the value of the major one. As for the methods of magnetic confinement employed, both toroidal and poloidal fields are generated by the plasma itself, while a steady magnetic-bottle field is provided by external coils (see fig. 46). This is possible since a spheromak supports the flow of an axial current through a field-reversed θ-pinch plasma.

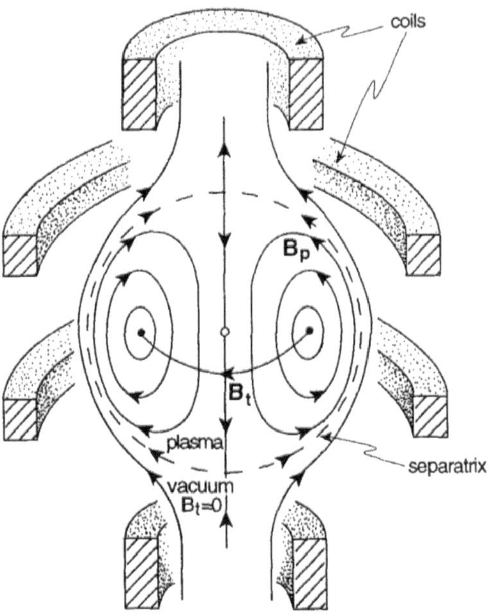


Figure 46: Poloidal and Toroidal fields of a spheromak

Topologically, the concept can be summarised as having open confinement systems, with the plasma being contained within a closed separatrix surface. Not unlike a reversed field pinch, the region inside the separatrix has a safety factor less than one. Moreover, this value is zero near the boundary of the plasma. The main advantage of spheromaks is their geometric compactness and simplicity. Experiments of such devices have only been operated for short pulse lengths. Most pulsed systems suffer due to the inability of most periods to contain the plasma in a stable manner. In addition, they experience relatively large radiation power losses due to the impurities released by the intense interaction of the plasma with the surrounding wall. Thus, until spheromaks are able to assume a steady-state operation, or are able to in some way compensate for the immense losses mentioned above, they remain, along with other alternative concepts, less promising for fusion as compared to tokamaks and stellarators.

## References


Aasen, A. *Nuclear Reactors, Nuclear Fusion and Fusion Engineering.* 2013. Accessed 29 Jan. 2023.

Arnoux, Robert. "'Proyecto Huemul:' The Prank That Started It All." *ITER*, 26 Oct. 2011, www.iter.org/newsline/196/930. Accessed 29 Jan. 2023.

Chen, Francis F. *Introduction to Plasma Physics and Controlled Fusion.* 3rd ed., Springer, 2016.

Clery, Daniel. "JET Fusion Reactor Passes 30 and Plunges into Midlife Crisis." *Science*, 12 July 2013, www.science.org/doi/10.1126/science.341.6142.121. Accessed 29 Jan. 2023.

Coblentz, Leban. "The Tao of Q." *ITER*, 30 Oct. 2017, www.iter.org/newsline/-/2845. Accessed 29 Jan. 2023.

"Collisionality Driven Turbulent Particle Transport Changes in DIII-D H-mode Plasmas." *OSTI.GOV*, 6 May 2020, www.osti.gov/biblio/1618050. Accessed 29 Jan. 2023.

"DIII-D National Fusion Facility." *General Atomics*, 3 June 2020, www.ga.com/magnetic-fusion/diii-d. Accessed 29 Jan. 2023.

E., Richard, et al. *Fundamentals of Nuclear Science and Engineering Third Edition.* E-book ed., 2016.

"EAST- Experimental Advanced Superconducting Tokamak." *Institute of Plasma Physics- Chinese Academy of Sciences*, 15 May 2015, english.ipp.cas.cn/rh/east/. Accessed 29 Jan. 2023.

*Fusion Power Explained – Future or Failure*. Produced by Philipp Dettmer, 2016.

Harms, A. A., et al. *Principles of Fusion Energy.* World Scientific, 2000.

"History of Fusion." *EUROfusion*, 11 Oct. 2021, euro-fusion.org/fusion/history-of-fusion/. Accessed 29 Jan. 2023.

"JET Makes History, Again." *ITER*, 14 Feb. 2022, www.iter.org/newsline/-/3722. Accessed 29 Jan. 2023.

"JET: The Joint European Torus." *Culham Centre for Fusion Energy*, 7 Nov. 2021, ccfe.ukaea.uk/research/joint-european-torus/. Accessed 29 Jan. 2023.

Kelly, Éanna. "The Joint European Torus Is Going out with a Bang." *Science Buisiness*, 28 Feb. 2019, sciencebusiness.net/news/joint-european-torus-going-out-bang. Accessed 29 Jan. 2023.

Kembleton, Richard. "Nuclear Fusion: What of the Future?" *ScienceDirect*, 16 Nov. 2018, www-pub.iaea.org/mtcd/publications/pdf/csp_019c/pdf/ov1_4.pdf. Accessed 29 Jan. 2023.

Knoepfel, H. *Tokamak Reactors for Breakeven: A Critical Study of the Near-Term Fusion Reactor Program*. 1978. Accessed 29 Jan. 2023.

Konings, R.J.M. *Comprehensive Nuclear Materials, Volume 4: Radiation Effects in Structural and Functional Materials*. E-book ed., 2012.

---. *Comprehensive Nuclear Materials, Volume 2: Material Properties / Oxide Fuels for Light Water Reactors*. E-book ed., 2012.

Lazarus, E. A., and G. A. Navratil. "Higher Fusion Power Gain with Profile Control in DIII-D Tokamak Plasmas." *eScholarship*, 18 Nov. 2021, escholarship.org/content/qt8g59w2zh/qt8g59w2zh_noSplash_c1073934597ff5ec46a0379f4bfa6dbf.pdf?t=p15qnd. Accessed 29 Jan. 2023.

Palumbo, D., et al. *Muon-Catalyzed Fusion and Fusion with Polarized Nuclei*. E-book ed., 1987.

Pamela, Jerome. *Overview of JET Results*. 20 Mar. 2003. *International Atomic Energy Agency*, www-pub.iaea.org/mtcd/publications/pdf/csp_019c/pdf/ov1_4.pdf. Accessed 29 Jan. 2023.

"60 Years of Progress." *ITER*, 5 Aug. 2021, www.iter.org/sci/BeyondITER. Accessed 29 Jan. 2023.

Tanabe, Tetsuo. *Tritium: Fuel of Fusion Reactors*. E-book ed., 2016.

Woods, Leslie Colin. *Theory of Tokamak Transport: New Aspects for Nuclear Fusion Reactor Design*. E-book ed., 2006.

Xu, Yuhong. *A General Comparison between Tokamak and Stellarator Plasmas*. 31 Mar. 2016. *Matter and Radiation at Extremes*, aip.scitation.org/doi/10.1016/j.mre.2016.07.001. Accessed 29 Jan. 2023.

Zhenqiang, Huang, and Huang Yuxiang. *Cold Nuclear Fusion Reactor and New Modern Physics*. 2013, Cold nuclear fusion reactor And new modern physics. Accessed 29 Jan. 2023.

Zohm, Hartmut. *Magnetohydrodynamic Stability of Tokamaks*. E-book ed., 2015.

# NOTES

# NOTES

# NOTES

# NOTES

www.ingramcontent.com/pod-product-compliance
Ingram Content Group UK Ltd.
Pitfield, Milton Keynes, MK11 3LW, UK
UKHW042008190726
13854UKWH00005B/2209

9 788119 510283